Environmental Toxicology, Economics and Institutions

Economy & Environment

VOLUME 8

The titles published in this series are listed at the end of this volume.

Environmental Toxicology, Economics and Institutions

The Atrazine Case Study

edited by

Lars Bergman
Stockholm School of Economics,
Department of Economics,
Stockholm, Sweden

and

D. Michael Pugh
University College Dublin,
Veterinary Faculty,
Dublin, Ireland

SPRINGER-SCIENCE+BUSINESS MEDIA, B.V.

A C.I.P. Catalogue record for this book is available from the Library of Congress.

ISBN 978-94-010-4419-6 ISBN 978-94-011-0968-0 (eBook)
DOI 10.1007/978-94-011-0968-0

Printed on acid-free paper

First published 1994
Reprinted 1995
02-0495-250-ts

The **European Science Foundation** is an association of its 55 member research councils, academies, and institutions devoted to basic scientific research in 20 countries. The ESF assists its Member Organisations in two main ways: by bringing scientists together in its Scientific Programmes, Networks and European Research Conferences, to work on topics of common concern; and through the joint study of issues of strategic importance in European science policy.

The scientific work sponsored by ESF includes basic research in the natural and technical sciences, the medical and biosciences, the humanities and social sciences.

The ESF maintains close relations with other scientific institutions within and outside Europe. By its activities, ESF adds value by cooperation and coordination across national frontiers and endeavours, offers expert scientific advice on strategic issues, and provides the European forum for fundamental science.

This volume records the efforts of Working Group 4, "Economic Analysis for Environmental Toxicology", of the Environmental Toxicology Programme and its daughter programme Environmental Damage and its Assessment. The activity served also a pilot study for the programme Environment, Science and Society: An Economic and Institutional Analysis.

Further information on ESF activities can be obtained from:

European Science Foundation
1 quai Lezay-Marnésia
F-67080 Strasbourg Cedex
France

Tel: (+33) 88 76 71 00
Fax: (+33) 88 37 05 32

TABLE OF CONTENTS

Acknowledgements

This volume exists because of all those whose initial, occasional, episodic or constant help, contributions, encouragement or diplomatic skills enabled this difficult, pioneering, inter disciplinary project to begin, to continue and, finally, to produce. To name all who played a part in each of the strands is not attempted, but to list the members of Working Group 4 is to record the interpersonal and interdisciplinary bridges built between its members and within its subgroups as an achievement of this ESF programme which would otherwise have gone unrecorded.

Working Group 4

Economist	Prof. Lars Bergman	Sweden
	Prof. Per-Olov Johansson	Sweden
	Dr. Tore Söderqvist	Sweden
Economists/ Institutionalists	Dr. T.M. Swanson	United Kingdom
	Prof. dr. Michael G. Fauré	The Netherlands
Toxicologist	Dr. Richard Lloyd	United Kingdom
	Dr. D. Michael Pugh (Chairman)	Ireland
	Prof. Marco Vighi	Italy
Former Chairmen	Prof. dr. J.B. Opschoor	The Netherlands
	Prof. dr. Harrie Govers	The Netherlands
ESF Secretariat	Dr. Jan-Henrick Kock	France

1.THE ORIGINS AND AIMS OF THE EUROPEAN SCIENCE FOUNDATION JOINT PROJECT IN ENVIRONMENTAL TOXICOLOGY, ECONOMICS AND INSTITUTIONS.

D. MICHAEL PUGH
Department of Small Animal Clinical Studies,
Veterinary Faculty,
University College Dublin,
IRELAND.

1. The European Science Foundation and Toxicology

The European Science Foundation (ESF) is an association of 55 member research councils, academies and institutions devoted to basic scientific research in 20 countries [1]. With headquarters in Strasbourg, it was founded in 1974 and sprang from a decision taken at a meeting of the West European Research Councils, Aarhuus, 1972. An intention then, and now, was to build "a little bit of Europe through the development of a coherent community of science and scholarship" [2]. To achieve this the ESF "brings European scientists together to work on topics of common concern, to co-ordinate the use of expensive facilities, and to discuss and define new endeavours that will benefit from a co-operative approach". That science which is "sponsored by ESF includes basic research in the natural sciences, the medical and biosciences, the humanities and the social sciences". Taking a broader view of Europe than that defined by certain well-known political and military organisations, the ESF has lived up to its claim to "link the scholarship and research supported by its members and add value by co-operation across national frontiers" [1]. The inter-disciplinary work reported in this volume well exemplifies this claim in having found Scandinavian economists to assist with a problem characterised by Italian biologists as part of a larger programme whose long-term goal is to provide to those who regulate for chemical safety world-wide an extended and improved database for future decision-taking. The regulatory interest arose with the EC Commission and has been included in the study with inputs from the Netherlands and the UK. The recent events in the former Eastern Bloc Countries and the revelation of the extent of environmental degradation in those territories make this study into drinking water contamination especially timely. They underline too the value of the long-standing interaction enjoyed by the ESF with its Eastern members and through its support of study visits to Western countries.

In addition to its fellowship programmes, the "connecting abilities" of the ESF are expressed via the establishment of networks of researchers, now 25, and the organising of *ad hoc* groups or workshops. The areas chosen for research co-ordination are specified in programmes, or "additional activities" and are adopted via a committee process culminating in the annual November meeting of the General Assembly of the ESF.

L. Bergman and D.M. Pugh (eds.), Environmental Toxicology, Economics and Institutions, **1-14.**

In this way the Environmental Toxicology (ET) Programme was adopted in 1989 for a five year term. It had been designed by a Scientific Advisory Committee for Environmental Toxicology (SACET) established during 1987. Toxicology had been a concern given effect as early in ESF history as 1975 but, reflecting the early dominance of the European Medical Research Councils, was concerned mainly with human health problems. It 1983 a Working Group in Environmental Toxicology (WGET) was established, initially to report on selected chemical substances under contract to the CEC. Responding to pressure to deal with more fundamental aspects of environmental toxicology, SACET was created to devise a strategy for moving the ESF focus away from what was felt to be the sufficiently provided-for study of mammalian, species-specific toxicology towards the much less well understood area of predicting the potential threat to man from substances present in his environment. SACET responded by suggesting that a European Research Institute be established, that an ambitious field project be undertaken and that the Programme of Grants in Toxicology (PGT), begun in 1982, be expanded to fund research in environmental as well as mammalian toxicology. While the latter proposal was accommodated within the Programme for Fellowships in Toxicology (PFT) which succeeded the long-running PGT programme in 1991, the grander proposals were beyond ESF resources or ambitions.

The infant states of ecotoxicology (the study of foreign chemicals on ecosystems) and environmental toxicology (the study of harmful effects of the environment on man) in comparison to the ambitions of those who wished to protect the environment and wished to use the findings of ecotoxicology as a basis for devising enforceable protective measures, caused the committee many problems. These were reflected in the two years which elapsed between the founding of SACET and the completion of its proposal for an Environmental Toxicology Programme.

2. The Environmental Toxicology Programme

The following text is the background material from the submission which lead the General Assembly of the ESF on November 1987 to provide the means which enabled SACET to undertake the detailed planning of the ET Programme. It is included because it describes the means by which chemicals come to be present in the environment, why their presence can threaten life-forms and how this threat can be identified, quantified and managed.

2.1. THE MANAGEMENT OF ENVIRONMENTAL RISK FROM CHEMICALS

The aim of any environmental protection policy is essentially one of managing the risks to human health and to ecosystems resulting from the impact of human activities on the environment. Risk management must be based on sound risk assessment, i.e. an objective and scientific evaluation of the environmental consequences of human actions. A major impact on the environment is that of chemicals which are released in ever larger quantities and larger numbers. These may be synthetic chemicals (xenobiotics) or natural chemicals which are cycled in larger fluxes through human actions. The study of these chemicals in the environment and their impact on man, other living organisms and the ecosystems includes in a logical sequence the following steps:

(1) Sources and releases, which include:

(a) Commercial products and their impurities which are either voluntarily released (e.g. pesticides) in the environment or involuntarily disseminated (e.g. PCBs);

(b) The various effluent streams, released into the atmosphere or into water bodies, originating from combustion, industrial processes, bio-industries, households, etc.;

(c) Solid or liquid wastes, urban or industrial, which are deliberately dumped.

(2) Once released in the environment these chemicals are transported and subject to physical and chemical transformations. Thus the study of the fate of chemicals in the environment includes such things as diffusion models, processes of abiotic (hydrolysis, photolysis, etc.) and biotic transformation or degradation, the accumulation and bioconcentration in the food chain leading possibly to man.

(3) The chemicals may then be taken up by organisms through inhalation, ingestion or penetration through membranes. As a consequence a certain amount of the substances, the dose, is accumulated in the organisms and possibly concentrated in some critical organ. The estimation of the exposure and dose are two of the main problems of environmental toxicology.

(4) The following step is that of elucidating the dose-effect relationships. Effects may be acute, sub-acute, sub-clinical, chronic, somatic or genotoxic, etc. The kinetics and toxicodynamics of chemicals in man and in non-human organisms may be similar or may differ, and this in itself is a subject for study.

(5) With regard to non-human targets the effects must be studied not only at the level of the individual but also at the level of the population, the biological community (biocenosis), and the whole ecosystem. In fact, while for human or domestic animals the well-being of the individual is the main concern, for ecological systems what is important is the ability of populations to maintain themselves, the preservation of the balance of the interactions among populations and the unimpaired functioning of the whole system.

(6) A detailed study of these response relationships leads to the establishment of environmental quality criteria (risk characterisation). These constitute the scientific input into policy measures which must also take into account socio-economic considerations and such things as cost benefit analysis.

(7) Once regulations or other policy measures for managing the risk (i.e. leading to its reduction) have been implemented, their effectiveness must be verified - hence the need for monitoring systems based on sound methodologies. A careful surveillance of the environment will establish whether the observed damage is being reduced or whether unsuspected negative effects appear, an outcome which may lead to a revision of the regulations.

The relationship between these various components of environmental policy may be represented schematically in the following figure.

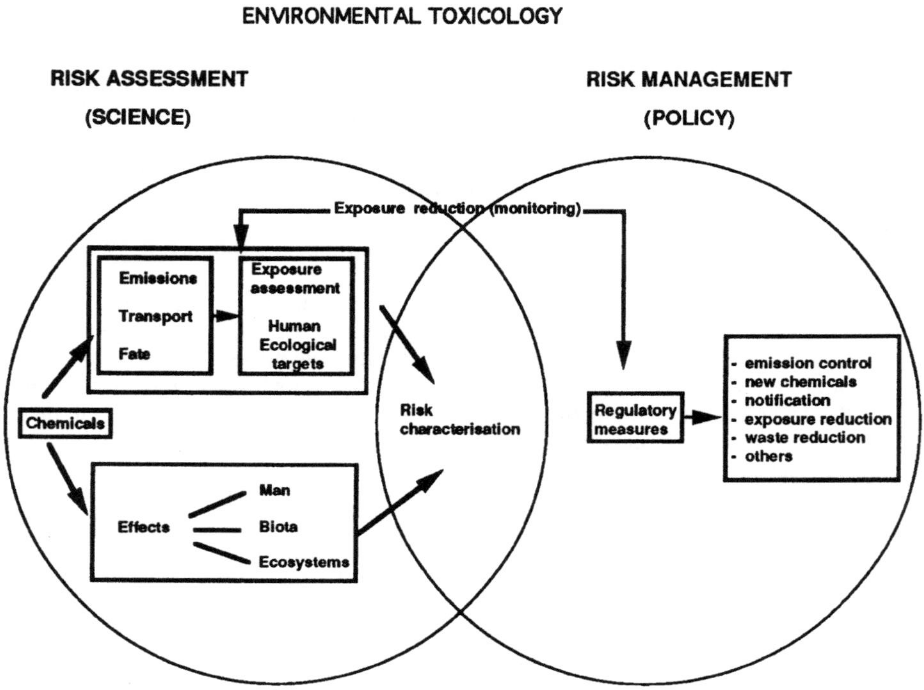

Figure 1. Environmental toxicology is concerned with chemicals in the environment as they affect human health, non-human biota and ecosystems. It produces assessments of risk (i.e. probability of damage), which are the scientific basis of risk management (i.e. environmental protection policy).

While many organisations and agencies deal with risk assessment it was felt that the ESF could play a useful role in line with its charter. It should concentrate on the scientific principles which are generally applicable to guide authorities and industry in taking proper measures to protect the environment. Of special interest in this respect are the improvement of methodologies to assign priorities for control of chemicals, as well as for the prediction of the total inter-media environmental impact of a substance. An ESF action could also provide guidance to the scientific community in identifying research priorities in environmental toxicology. In line with the above-mentioned considerations the advisory committee identified a number of research areas which could be tackled by the ESF. Some of these are rather broad and others much more specific [4].

During 1988, SACET refined its positions on several components of environmental toxicology where further work was necessary for the growth of the discipline and for improving its utility to the regulatory fraternity. The ET programme as proposed and adopted specified the following four broad areas in each of which progress was considered both possible and desirable for environmental toxicology itself and for its ability to produce data of greater utility in environmental management policy and practice.

2.1.1. Project 1:Chemical Exposure Prediction
An integrated approach using models of environmental distribution, laboratory simulation and selected field data should be the basis for predicting potential exposure for humans and for ecosystems. Early warning systems for the assessment of exposure can be of relevance to regulatory agencies for the management of chemicals in the environment.

2.1.2. Project 2: Assessment of Environmental Damage
There is a need to discuss and develop methodologies for the assessment of environmental damage and their relative importance. In particular the group must evaluate those methods which are able to identify changes in ecosystems as well as the early-warning systems which may be able to predict potential irreversible changes in populations or communities.

2.1.3. Project 3: Risk Assessment Methodology for Environmental Effects of Chemicals
There is only limited international homogeneity in the evaluation of the environmental impact of chemicals. There is, therefore, a need to compare the various systems for the assessment of risk of environmental damage.

2.1.4. Project 4: Economic Analysis for Environmental Toxicology
Economists have had for a long time a strong interest in the analysis of environmental issues. For the cases of acidification, eutrophication and noise, for example, numerous studies have been carried out dealing with the cause, the effects and the alternative ways of resolving these issues. The applications of these methods to persistent chemicals creates a new challenge for environmental economists. In particular economic analysis provides possibilities for setting priorities [5].

2.2 ENVIRONMENT, SCIENCE AND SOCIETY

A programme of this name, under the remit of the Social Sciences Committee of the ESF was begun in 1990. Like the ET programme, it operates by means of small working groups and research projects. Of particular interest to the ET Working Group for Project 4 (ETWG4) was that the main fields of enquiry of the Environment, Science and Society (ESS) Programme were to be in environmental economics and the institutional response. By 1990 SACET had already conducted a seminal workshop "Economic Analysis for Environmental Toxicology: Applications to Persistent Micro-Pollutants" to explore the interface between environmental toxicology and environmental economics and the utility of both to institutionalists [6]. This obvious confluence of interests between the ET and ESS programmes lead to a rationalisation of effort and it was agreed the "Economic Analysis for Environmental Toxicology" strand of ET be carried forward as a joint, inter-disciplinary, research activity between the two programmes. The vehicle for such an exploratory co-operation, as recommended in a report on the May 1989 Noordwijkerhout Workshop, should be a case study [7]. Within this the pathways of hazard, identified by natural science investigative techniques, will be subject to economic and regulatory scrutiny and jointly evaluated [8] so as to identify those areas, if any, in which co-operation could be fruitful.

It should at this point be admitted that there were those among the economists at Noordwijkerhout who felt that the time was not yet ripe for a fruitful interaction with ecotoxicologists, and not only because much toxicological data was perceived as not constituting a basis adequate for predicting or valuing environmental degradation. Similarly among the toxicologists there were those who felt that the concern of environmental economists to devise ways of taxing pollution indicated a greater interest in generating revenue than devising an economy which ensures that any development will be biologically sustainable. It is, of course, in just these circumstances that the ESF is challenged to seek progress and the continued existence of ETWG4 some four years later bears witness to the skill and determination of ESF staff. Indeed, its commitment to seeking progress at the interfaces of toxicology, economics and institutions was emphasised further when the General Assembly of the ESF agreed to provide specific additional funding via a new programme "Environmental Damage and its Assessment (EDA) in November 1990 [9].

The two following sections describe the rationale for the composition of ETWG4 and the route by which the group selected a problem suitable for interdisciplinary study.

3. Why Economic Analysis for Environmental Toxicology

The Environmental Toxicology Programme was designed almost entirely by toxicologists for toxicology and this is evident from the four specific activities proposed. These were:

(1) Chemical Exposure Prediction;
(2) Assessment of Environmental Damage;
(3) Risk Assessment for Environmental Effects;
(4) Economic Analysis for Environmental Toxicology.

The first three topics are common to all attempts to manage health risks from chemicals be they from industrial chemicals to workers, from medicines to patients or from pesticides to non-target species. The central belief of all toxicologists is that damage from chemicals will be seen in any life-form when it encounters the chemical in sufficient quantity, i.e. the effect categorised as damage is dose or concentration-dependent. Safety, the usually desired state, is achieved by ensuring that the life-form is never exposed to a dose or concentration sufficient to cause damage. This means that in the safe management of chemicals the first requirement is to identify the threshold dose or concentration which just fails to elicit the adverse response, the so-called No Adverse Effect Level or NAEL. After that a management practice designed to ensure that sensitive populations are never exposed to the chemical at amounts or concentrations greater than the NAEL will provide an acceptable or tolerable level of exposure and hence, safety. To put these principles into practice requires, in turn, various scientific skills. These include the ability to be able to measure and/or predict accurately the dose or concentration of the chemical of interest to which the life-form becomes exposed. Secondly, it is essential that all of the changes which a chemical can cause to a life-form be known and that a distinction be made between those which are mere adjustments to the presence of the chemical and those which are regarded as harmful. Armed with these pieces of strictly scientific information it is then possible to predict the extent to which, in our case, the environment will be damaged following an exposure to the chemical of a particular duration and at a particular concentration.

To improve practices in these areas was the stated goal of SACET and this is given effect through the first three specific activities listed above.

Toxicologists are applied biologists and would in the main react conservatively to the detection of chemical damage in that their instinctive reaction, whether as doctors treating poisoned patients or ecotoxicologists surveying a fish-kill in a polluted river, would be that such happenings should be prevented and that there should be in place laws which outlaw the circumstances which gave rise to the damage. It is, perhaps, a reflection of their biological ideals that so many would think so.

The management of risks is in reality subject to other considerations, however. In our material world defence of well-being and quality of life depends on the health of national economies. In consequence a more tempered view of safety management usually prevails.

The chemical which causes the poisoning can often be one capable of delivering economic benefits to at least some sections of the community. The production of crops is simplified and made more profitable to farmers when appropriate weedkillers or insecticides are applied. The cost of the produce to the consumer is in consequence lower and supplies are more plentiful and more regular. Those who manage risks do so by weighing risk against benefit in a quantitative sense. Benefit is of course seen on dimensions other than food production. More general socio-economic and political inputs often figure in the calculus and it is here, for example, that economists can make a valued contribution. Indeed, devising incentives to up or down regulate demand is a standard approach to influencing the size of a market for a particular commodity. The end-point is a decision in which a certain price (an element of risk) is considered acceptable in relation to the benefits which then become available. The skill of the manager is reflected in the proportion of the total population which finds the compromise acceptable.

The task addressed by the fourth specific activity therefore, gives effect to this real world view of the risks from chemicals. The scope of the study in its narrowest sense could have been to seek an economist's input into deciding the scale (and hence cost) of the toxicological studies which a manufacturer should conduct before the chemical product is marketed. An alternate interest could be to look at the cost implications of predictive toxicology versus those of an environmental clean-up, should an untested product be shown to be damaging only after it has been marketed. A third valuable component will be that of bringing home to toxicologists the reality that control or management of chemicals is a practice which is influenced heavily by considerations other than their toxicity profiles. Indeed, it is a practice which is heavily influenced by economic analyses, especially those which promise, for example, the prevention or reversal of unemployment. By pooling their assessments, it is possible that better managemental practices, lying somewhere between the banning of synthetic chemicals and a truly laissez-faire approach to markets, might be evolved. For example, a toxicologist would identify an efficient amount of a pesticide in the environment as that which delivers agricultural benefit without adverse effect on other life forms exposed inadvertently to the substance. The economist's vision of efficiency would depend on the effect of a supply and demand relationship on the amount of the chemical actually used by the farmer. Conceivably the two estimates of an efficient amount to use could be made to converge if to the supply and demand interaction were to be added a factor which adds a concern for environmental quality.

As yet nothing has been said of the role of institutionalists in the team assembled to give effect to the topic of Economic Analysis for Environmental Toxicology. Institutionalists analyse and help to shape the means by which those who manage populations and economies exert their influences. Principally, this refers to the legislative measures produced by governments. The institutionalist, then, often a lawyer, is well placed to integrate or assess the consequences of inputs from interest groups into the political process. Toxicologists and economists are just two of these groups. An ability to weigh the quality of their small in number but highly focused inputs against the more diffuse and frequent inputs of vociferous pressure-groups, media presenters or even against a perception of voter-preferences would plainly be a valuable attribute of the institutionalist and the politician.

4. Approaching Collaboration

In making its proposals for the development of environmental toxicology, SACET had identified four areas within the discipline where progress could be made. These had in common an opportunity or even a requirement for interdisciplinary collaboration.

While research at discipline interfaces has often been particularly productive it does of course have its peculiar characteristics. One such is the initial need to come to understand enough of the philosophies and methods of the novel discipline or disciplines so as to be able to frame cross-discipline questions worth the answering. Having succeeded at this level, next came the need then to restrict intellectual curiosity to the extent necessary to remain within the confines of the Environmental Toxicology Programme. Finally, the topic for study would have to be one which would allow balanced inputs from the natural scientists, economists and lawyers who made up the working group. The remainder of this section deals with the steps by which the group specified a research topic appropriate to its task and to its composition.

The obvious ground common to all ETWG4 members was an interest in environmental change and an acceptance that any such change would create consequential costs of some type. The natural scientists could describe and quantify change in biological terms, the economists could assess the extent to which the change distressed or deprived man in monetary terms while the institutionalists could devise means for regulating the rate of change, perhaps using information generated by science or economics. The ultimate choice of topic would plainly have to allow for a proper intersection of those interests if it were to serve its purpose as a pilot study in which to look for and perhaps evaluate the benefits of collaboration.

Remembering that a major reason for the existence of the ET programme was the acceptance of the view that environments do become damaged, that this damage can readily pose risks of adverse health effects to man and other biota, that the use of chemical substances can damage both man and the environment, that the use of chemicals is highly regulated and that a goal of the programme was to provide bases for the taking of better regulatory decisions in the future, it was perhaps inescapable that the area of intersection should contain an environmental pollutant which originated in the chemical industry.

Toxicology is concerned with identifying the adverse effects of chemical substances on life forms and, by means of dose-response studies, with quantifying the concentration-dependent abilities of the substances to produce damage. By means of such studies modern toxicology contributes to the prevention of chemical disasters. This it achieves by attempting to predict the exposure circumstances which would in all probability be unlikely to cause harm. Economists could value any damage done and advise on the economically efficient amount of precaution to be taken (in pre-marketing predictive toxicology studies) before the substance is released into the environment (i.e. as a consequence of marketing, use and/or disposal). In this way economists can achieve internalisation by the industry of the externalised cost which arises as environmental damage. The industry thus assumes not just the production costs but the true total costs arising from the product. Finally, the institutional interests would be well served by studying a substance already regulated at the time of market entry, but which has since been in breach of those regulations. This would provide for an analysis of the "failure" of the existing regulations and the proposal of alternate regulatory scenarios which could perhaps have prevented the breach.

Because the case was to be a pilot study, it was also judged to be desirable to select in favour of simplicity. For example, pollution is proven only when chemical analysis reveals the presence of the xenobiotic (any chemical foreign to the environment in which it is found) in the sample tested. Such analysis is simpler when extraction of the chemical from the sample can be rapid, largely complete and with minimal accompanying contamination. These circumstances are most closely approached when the sample is water and even more so where, apart from the pollutant of interest, the water is but lightly contaminated such as that pure enough for human consumption. The first decision of the group was to adopt as its project "The Environmental Impact of Regulated Chemicals on Water Uses".

The purity of water for consumption by man is defined by Directive 80/788 EEC and this has been given effect in all member states. This common standard taken together with a chemical entity known to be or nearly to be in breach of the standard defined by the Directive in several member states, would give the chosen topic a European relevance appropriate to ESF aims.

If such a chemical simply did not exist in nature, then its presence in drinking water would be due entirely to its industrial production and escape into the environment. If it were a product and had a single, specific use, then not only would its route of entry into the environment be known, but also accessible would be the quantity used in the target compartment which resulted in the appearance of the measured, pollutant concentration in drinking water. If the pollutant arose from a commercial product then it should be possible, with industrial assistance, to obtain retrospective information to establish both the pattern and the quantity of use, the production and the regulation costs.

If the substance had been regulated at market entry and had belonged to a product class which required toxicological information to be gathered and assessed pre-marketing, such as a pesticide, then there could be a presumption that its presence in drinking water at a concentration in breach of that regulation was undesirable for health reasons. Such a substance would satisfy toxicologists, institutionalists and economists and, if banned, would allow a full before (*ex ante*), during market life and after withdrawal from the market (*ex post*) analysis. "The Impact of Pesticides on Water Uses" was for these reasons selected as the more restricted topic needed for the pilot study. It remained only to select a pesticide whose physicochemical properties would be compatible with a tendency to enter and persist in water, viz. solubility in water, low volatility, low tendency to bind to soil constituents and resistance to degradation by photolysis, microbial attack and hydrolysis. Perhaps surprisingly, a substance which satisfied all of these and the previously mentioned criteria was known to exist. The sale and use of atrazine, a selective herbicide used for weed control in agriculture, were banned in Italy during 1990 because it had exceeded the maximum concentration threshold for a single pesticidal substance in ground and hence well water. The legal obligation to monitor water quality would mean that records which would allow a study of the effect of application rates of atrazine on its presence in drinking water over several years should exist. This database would allow the effect of the subsequent ban also to be studied. The post-ban phase could include also a survey of replacement herbicides in a comparative study and a survey of the costs/benefits arising from any pollution abatement measures.

The expectation that the data-base for atrazine would be largely complete with respect to toxicology and institutional components and readily completable in the case of ecotoxicology and economics promised an adequate basis for testing of both the design and the operation of the regulatory measures in place throughout its commercial life.

A study using atrazine as a focus for the intersecting interests of toxicologists, economists and institutionalists would of necessity be a retrospective analysis of the control system in place throughout the already-ended commercial life of the banned substance. While this could of course be held to limit that scope for theory development which would exist in a wholly unconfined scenario, it does have considerable relevance and advantage. Not the least among these is the need to recognise that the project is a pilot study also within the context of the Environment, Science and Society Programme and as such has to conclude rapidly if it is to fulfil that role with benefit. In describing that programme, the chairman of its steering committee, Professor T. O'Riordan, commented on the rarity of major interdisciplinary studies of complex environmental systems involving both the natural and the social sciences. He went on to note that the combination of theory building and theory testing through hands-on case studies is even rarer [8].

It was its ready appreciation of the relevance of the proposed topic for study to its wishes to seek developments in environmental toxicology which enabled SACET to adopt the atrazine case-study on behalf of the Environmental Toxicology Programme. It was its recognition that this study offered a unique opportunity for an appraisal of the benefit which might arise from a co-ordination of the differing inputs from toxicologists, economists and institutionalists in a truly novel area which lead SACET later to propose for it the separate and additional funding via the Environmental Damage and its Assessment Programme, which, as mentioned previously, was adopted at the ESF General Assembly in November, 1990.

5. The Database and the Analysis

This chapter has described the framework which gave rise to a decision to establish a group to explore the topic of economic analysis for environmental toxicology. It described also the rationale for the composition of the group and the route by which the group selected a problem suitable for interdisciplinary study. It is completed by an account of approaches taken by the toxicologist, economist and institutionalist members of ETWG4 in their attempts to collaborate and interact fruitfully within the programme of study.

It was recognised long ago that man's activities may damage the environment. The tendency identified in the pioneering work of Marsh in the middle of the last century was that deforestation, drainage or irrigation could upset the balance of nature. In the atrazine case, it is the pollution of groundwater by a pesticide which exemplifies his belief that "while man cannot destroy elementary matters. he has the power irreparably to damage the combinations of inorganic matter and organic life" [10]. This first chapter has described how ETWG4 selected the exceedence of a drinking water quality limit by atrazine as a model case upon which to study the forces which lead to the transgressed legal instrument having the form that it did, in which to study the failure of the instrument to prevent man from damaging drinking water and through which to develop alternate strategies for the future.

The second chapter introduces the reader to those properties of chemicals which give rise to health concerns and which, in consequence, fuel the drive to regulate chemicals on the basis of their experimentally-revealed capabilities to cause adverse effects. Emphasis is given to the possibility of being able to legislate with greater severity according to the scale of risk, e.g. based on amounts entering the environment, the tendency to persist within the environment or the tendency to accumulate within a life-form. The failure of simple market forces adequately to regulate the rate of entry of chemicals into the environment is explained, as is the failure of the drinking water directive to return the costs of environmental pollution directly to those who benefited from the sale and use of the chemical which then caused the pollution. This unfairness could be corrected, for example by exerting a Pigouvian tax on the use of pesticides. This would either regulate pesticide use down to an acceptable level or would generate the fund necessary to correct any damage to the environment which might follow its use, i.e. the tax would convert the external cost into an internalised cost.

At the practical level it is questionable whether sufficient certainty exists about the adverse properties of pesticides, their distribution through the environment or their uptake by man to allow for a satisfactory estimation of the necessary user tax. The existence of these uncertainties poses for the economist problems in advising on the cost-benefit consequences of wishing to reduce them, e.g. by determining the scale of ex-ante (pre-marketing) testing of the pesticide. How big a test is big enough? How often should the test be repeated? How appropriate is the test as a predictor of a human health risk? An analysis of a study of 64 tests which could predict carcinogencity when applied to the same set of chemicals showed that the law of diminishing returns applied to the cost of reducing this example of a reducible uncertainty.

The opposite circumstance applies to irreducible uncertainties, i.e. those which remain unchanged despite additional testing. This is exemplified by hazards which conventional testing would not reveal, as was the case for thalidomide 30-35 years ago. Two remaining sources of irreducible uncertainty arise from the need when attempting to predict, to extrapolate, e.g. from high dose to low dose exposure, or from rodents to man. The current approach to managing this problem is to put in place post-marketing surveillance schemes. This allows a verifying assessment of the predicted properties of the chemical with respect to human health and can reveal additional information, e.g. about one half of a drug's side effects in man are identified by this means.

While all of the foregoing addresses well the regulation of chemicals of any type by means of health-based criteria, the regulation of pesticide presence in drinking water would appear to have been driven by quite different concerns. The parameter value chosen equated well with the limit of sensitivity at which specialist chemical analysis was 20 years ago able to discriminate sufficiently confidently between the presence or absence of pesticides in drinking water. Plainly if a regulation defines the upper limit of acceptability of the presence of a chemical entity in water, then the lowest enforceable level at which that limit can be set is defined solely by the abilities of the analyst to distinguish the presence or absence of the chemical at that stated concentration.

The fact that routine chemical analysis can now achieve this discrimination very much more readily is often blamed for the embarrassing frequency with which drinking water supplies are found to be in violation of Directive 80/778 EEC, especially with respect to pesticides.

Similarly many argue that the limit should have been set with reference to health-based assessments of each individual pesticide, rather than at an overall value which was no more than a surrogate for zero. It is indeed the case that zero residues are very rarely used in regulatory practice for other contaminants of other dietary commodities, a somewhat higher health-based value nowadays being the norm.

Chapter 3 looks at the need for government to take action to protect or conserve the public good when supply and demand economics alone have failed to provide equity. Against this background the author goes on to describe the need for and content of Directive 80/778/EEC. In particular he stresses the absence of a stated rationale for the choice of the 0.1 μg/l maximal concentration threshold for single pesticides and an inconsistency in parameter values among the family of water quality directives. The mechanisms which could be used to regulate water quality are compared, namely the input control which seeks to prevent pollution versus the output control which requires corrective action if a threshold is exceeded. The sequence of actions taken by Italy in response to threshold exceedences by atrazine is recorded and discussed. The sanctions provided for were, in respect of their appropriateness and enforceability, felt to have been insufficient.

The data-base for the collaborative study on atrazine is provided by chapters 4, 5 and 6. The relevance of the problems experienced by Italy over atrazine to other EU member states is made plain by Chapter 4 which gives an account of the state of implementation of Directive 80/778/EEC via national laws throughout the community. These fail to internalise the cost of the damage to the public good of drinking water quality as, by their design, they are command and control instruments. Such instruments prescribe that action be taken only after, as was the case for atrazine, the legally-defined capacity of the environment to assimilate it had been exceeded. In Northern Italy this point was reached following the year-on-year accumulation of atrazine in ground water.

The Vighi and Zanin contribution (Chapter 5) looks into the properties of the environment which determine its capacity to assimilate the herbicides used in crop production. These include the ability of the environment chemically to degrade the foreign material into acceptable fragments and its ability alternately simply to render the foreign material non-available by binding it to soil components. In the absence of these two absolutes, the physicochemical properties of the herbicide, in relation to those of the soil and the deeper strata, as modified by water flow, will determine the time course of the leaching of the herbicide by water. Once in groundwater the absence of light and significant microbiological life contribute to the very long survival of the pesticide chemically intact, an issue of real concern because groundwater is often consumed as drinking water with minimal purification beforehand.

Atrazine and other pesticides have seen extensive use in maize production in the Po Valley of Northern Italy. It was in this region that atrazine exceedences were most frequently recorded and the region of Lombardia was chosen as the area in which the reported study was made. The chapter concludes with useful physico-chemical, agronomic, toxicological and environmental fate profiles of atrazine and four of its potential substitute herbicides.

Once groundwater contamination exceeds a legally specified concentration threshold, the options open to water authorities are limited. These include closing the contaminated supplies, blending the contaminated supplies with clean supplies so that the concentration of the contamination is sufficiently lowered or reducing or removing the offending contaminant. In Chapter 6, Nurizzo describes the processes by which decontamination of drinking water can be effected, either by producer or consumer. Usefully, the state of development of these processes is compared. Information is provided also on the flow rates and operational costs of installations of stated capacities. Under the present legislation, such costs fall on drinking water producers and consumers.

Plainly much remains to be done before decisions-takers approach the luxury of decision-taking from a fully-informed basis. For example there is a need to value various environmental attributes, goods and services. Only then can their degradation be costed and converted into, say, appropriate taxes. What of the capacity of the environment to assimilate? Should polluters pay according to the assimilative capacity fraction which their pollutant will occupy? Is a health-threat the only basis on which to set action thresholds? How does one accommodate consumer preferences, e.g. for herbicide-free drinking water? Is a preference defensible only if the consumer is prepared to pay for its attainment?

While Chapter 7 does not attempt to address all of these questions, the contributors from the Stockholm School of Economics in the concluding chapter of the book's analytical section evaluate the cost-effectiveness of two ways of meeting the atrazine ban in Northern Italy. This is approached by comparing the costs incurred by maize farmers in having to switch to replacement herbicides, i.e. input costs, with those output costs which arise from having to purify polluted water close to its point of consumption. While recognising that there would be many other costs arising from the atrazine ban, the authors conclude that the greatest cost is likely to be that incurred by the farmer in having to use perhaps more costly but less effective herbicides. As the Directive refers only to drinking water quality and not to the quality of the source from which it is obtained, it is economically valid to consider the costs of purifying drinking water. In effect this presumes that groundwater pollution is in itself acceptable provided that this has no adverse environmental effects. Data gaps plus the reality that atrazine is not the only pollutant in Alto Vicentino water made impossible the estimation of the true costing due to atrazine alone. That is perhaps a fine point when it was the atrazine concentration which triggered the need for the additional purification step. The filtration system would of course cope also with those herbicides which replaced atrazine so the total cost would continue to support a general heath benefit. By reference to WTP studies conducted elsewhere in the case of possible adverse health effect reduction in drinking water, the authors conclude that the Alto Vicentino consumers would probably be prepared to carry the increased cost of drinking water consequent on its increased production costs due to the atrazine ban in return for an improvement in the quality of the drinking water.

The concluding chapter, contributed by Bergman and Pugh, while recognising the extent to which the collaborative study has moved forward the debate over the issues which drive and shape environmental quality legislation, recognises also the large scale of the opportunities for further study. Nonetheless, Marsh's view that we cannot remain inactive until all of the necessary information has been produced "and exact science has taught it (the world) a better economy" is endorsed. The analysis of the Atrazine case has, however, indicated plainly the novel contribution which its three-pronged approach could make by posing the question of the desirability of using regulations not just to restrict, but with benefit to modify the motivations which drive manufacturers to specify preferred chemicals for development for particular market applications.

6. References

1. ESF Communications No. 27, p. 2, October 1992.
2. ESF Communications No. 21, P. 3, October 1989.
3. Item 7, *Draft Minutes of the Executive Council of ESF*, March, 1987.
4. Item 13, *Minutes of the General Assembly of the ESF*, November 1987,
5. Item 13a, *Minutes of the General Assembly of the ESF*, November 1988.
6. Opschoor, J.B. and Pearce, D.W. Eds (1991) *Persistent Pollutants: Economics and Policy*. Kluwer Academic Publishers, Dordrecht.
7. Opschoor, J.B. (1989) Report on ESF Workshop Economic Analysis for Environmental Toxicology: Applications to Persistent Micropollutants to SACET Meeting of July, 1989.
8. ESF Communications, No. 22, p. 18, April 1990.
9. Item 7.E, *Minutes of the General Assembly of the ESF*, November 1990.
10. Bingham, T.H. (1989) Social Values and Environmental Quality. In D.B. Botkin, M.F. Caswell, J.E. Estes and A.A. Orio (eds), Changing the Global Environment; Perspectives on Human Involvement, Academic Press, Inc., p.p. 367-382.

2. THE REGULATION OF CHEMICALS IN AGRICULTURAL PRODUCTION:

A Joint Economic and Toxicological Framework.

TIMOTHY M. SWANSON
Faculty of Economics
Cambridge University
UNITED KINGDOM.

RICHARD LLOYD
Formerly MAFF Fisheries Research,
Burnham-on-Crouch,
UNITED KINGDOM.

1. Introduction - Why are Synthetic Chemicals Regulated?

During this century the synthetic chemical industry has enjoyed rapid and continuous growth. By 1980, that industry manufactured £350 billion worth of chemicals world-wide; 50% of that occurred in Europe. The industry is now a giant in the European economy. It contributes 10% of valued added in most developed countries, 2-3% of GDP, and 10% of trade [1]. The phenomenal growth of the industry has seen a doubling in output volume every 7-8 years through its life (until 1980). Between 1950 and 1980, the amount of chemicals produced per capita in Europe more than doubled each decade [2].

The agricultural sector is one of the primary users of synthetic chemicals. In Europe the average annual application of synthetic fertiliser has increased from 123 kg/ha in 1965 to 229 kg/ha in 1985 [3]. The world production of pesticides is about 2.5 million tonnes/annum, and has been growing at about 12.5% per annum [4]. Two thirds of this production has historically been applied in Europe and America.[1]

Of course, the presence of chemicals in the environment is nothing new. In fact, the whole of the wider environment (air, water and earth) is chemical in nature. All tangible matter has both a physical and a chemical aspect, so that all that we experience in this world is literally of chemical origin be it natural or synthetic. In addition, human involvement in the environment has always altered the balance of chemicals that we ingest, through choices of what to eat and how to produce it.

For centuries, particular chemicals have been selected for specific purposes in order to provide certain benefits, e.g. to treat ailments, for colouring, or as preservatives.

[1]The current trend in chemical usage is towards stability in the developed countries, with massive increases in the amounts used in developing countries. This increased trade is not the subject of this paper, but it is a subject for concern. There is at present a movement to require the "prior informed consent" of a developing country's government prior to the export of any chemicals to it.

L. Bergman and D.M. Pugh (eds.), Environmental Toxicology, Economics and Institutions, 15-38.

For example, chemicals from fuchsia plants have been used for colouring in Ireland for centuries, nitrite curing (of pork) and fermentation (of milk products) has been used for preserving foods throughout Europe for decades or centuries respectively, and traditional medicines have developed in most places based on those chemicals of animal, vegetable or mineral origin available locally.

The novelty of the modern age lies in the process by which today's chemicals are formulated and manufactured. The past century witnessed the development of synthetic organic chemistry. Generally, this is the development of new chemical compounds by building from the simpler building blocks of organic chemistry, e.g. the hydrocarbon molecules refined from petroleum have been used to generate the monomers from which the familiar plastics are produced. It is this "building block" approach which distinguishes modern organic chemistry from the earlier use of chemicals [5].

It is this process of molecular construction and the range of properties of the resulting chemicals, which has produced the difference between current and previous periods of chemical usage. Chemicals with the desired properties are now built up from simple parts and processes, rather than relying solely upon (or intervening with) the natural world to generate the desired chemicals through biological processes. In fact, many of the chemicals now made by synthesis could be (and perhaps originally were) derived from a natural source. For example, several synthetic or semi- synthetic drugs have been derived and marketed as improved muscle-relaxants when compared with the original natural source of this valuable action, the South American arrow poison, curare.[2]

There are now synthetic substitutes identical to or substantially similar to nearly all naturally extracted drugs. The biological process for the production of a chemical is used only when it is the more cost-effective means of production; this is the case with regard to the manufacture of, for example, the drug penicillin which is produced through fermentation. Very often, however, production by synthesis is more cost-effective than natural production; also, the standardisation afforded allows for greater purity and lower cost quality control.

The modern chemical industry has resulted from this versatility and cost-effectiveness. The merit of synthesis of organic chemicals lies not only in its ease of replication and but also in its very independence from biological processes. This means that large numbers of chemical compounds could be generated and characterised in a matter of days or weeks rather than the months or years required to generate the same or similar compounds biologically, especially in nature.[3]

---

[2]This practice itself has local valuation implications for the environment, by denying the natural commodity its return and therefore reducing the value of natural habitat; however, this issue is outside of the scope of this paper.

[3]This is, of course, a discussion of the comparative advantage of organic chemical synthesis against the background of the nascent chemical age, i.e. prior to the advent of biological engineering. It is very likely that we are now entering into a new "biological age". The fact that there are similar concerns now about the impact of biological engineering on the environment indicates the general nature of the problem.

It follows that when a chemical with desirable properties is created a method for its mass production can be proposed almost immediately. With the additional step of formulation, large-scale production could be implemented almost immediately.[4] Large-scale replication of a laboratory process involves significant economies of scale, meaning that the unit cost of most bulk chemicals is quite low. If the market for the chemical is at all competitive, then this cost-effectiveness becomes a benefit for the consumer; for example, the cost of an application of pesticide to a fruit crop in the U.K. in 1979 was determined to be about £1.79 per ton of fruit.

In short, synthetic organic chemistry is another application of the principle of mass production in modern society.

This is the basis of a problem created by this industrial development - the sheer scale of the potential for the introduction of new synthetic chemicals often in large quantities and at low cost. Whereas the production of desired chemicals by natural or biological means implied an evolution of technology, over a period of time their synthetic production implies rather abrupt shifts.[5]

The problem of regulating the environmental safety of the synthetic organic chemical industry has much to do with formulating the societally appropriate testing procedures for its control. Gradual shifts in technologies allow for mid-course corrections while abrupt changes do not, due to the problem of irreversibilities. Therefore, much of this paper concerns the economics and practice of formulating and examining an optimal regulatory procedure in the context of abrupt changes.

Even with optimal testing and controls, however, there remains a core of ultimately irreducible uncertainty with regard to the reliability of this process, as there does with any process which involves expanding scientific and technological frontiers. This uncertainty regarding synthesised chemicals lies primarily in three areas: accumulation and persistence, synergisms and antagonisms, and human-specific effects. Dealing with irreducible uncertainties is therefore also an issue discussed in a later section.

The principal subject of this chapter is the development of a theoretical framework for the regulation of the synthetic chemical industry, especially those forms of synthetic chemicals which enter directly into the food chain - as additives or residues. The particular case to be examined will be the regulation of the agricultural chemical industry, with regard to the issues developed above. Although synthetic chemicals are a pervasive part of society, their main potential to exert a harmful effect on the general public is through their entry into the food chain. Therefore, we focus upon the health effects of certain chemicals on the general public (not those who incur special risks by virtue of working directly with chemicals) and the specific questions asked are (1) What should be the specific objective

[4]The synthesis of most organic chemicals is usually done in batches of about a thousand pounds, involving basic components and only a very small number of reactions [6].

[5]This is not the sole important difference. Another would be the difference in impurities in chemicals generated by the two processes. The impurities in synthesised chemicals are often residues of solvents and reagents utilised during their syntheses, biological impurities are usually much more complex, e.g. synthesis intermediates.

pursued in the regulation of these chemicals? (2) How can that objective be implemented effectively?

Initially these issues are pursued in a hypothetical state of "institutional vacuum", *i.e.* in the presumed absence of any regulatory structure. To begin, the type of health risks which might arise if chemicals were applied in agricultural production in the absence of national regulations are defined. Once those risks are identified, the nature of the need for a regulatory structure is then indicated. In conclusion an assessment is made of the extent to which existing regulatory institutions correspond to those perceived to be required.

2. Regulating the Risk from Synthetic Chemicals - The Mode of Inquiry

It is now common knowledge that some synthetic chemicals are potentially very dangerous. The well-known example of the teratologic adverse effects of the drug thalidomide brought home the potential for devastating consequences of the introduction to the market for human consumption of chemicals with previously unknown biological activities.

Sometimes unfamiliarity with the hazard and ignorance of the size of the risk contributes to its dreadedness. The inconceivably small doses required for the potential lethality of a few chemicals (less than 1 microgram/kilogram of body weight for highly toxic ones) causes understandable dread in a population where over a hundred million tonnes of chemicals are synthesised each year. Also unsettling was the discovery of the related phenomena of persistence and accumulation of some chemicals; for example, the organochlorine insecticides (DDT, aldrin, dieldrin) persisted in the food chain for years after their discontinuance, right to the point that human milk was potentially unfit for human consumption (according to WHO notional standards). Since there is little capacity for the lay person to distinguish between the relative potencies of the 63,000 synthetic chemicals in common use, it is not irrational that most people have come to perceive all chemicals as significant hazards to health; this has been termed chemophobia.

Nonetheless, the option of removing all chemicals from the environment does not exist; as discussed above, the environment is itself describable in chemical terms. The removal of all synthetic chemicals from the environment might be another approach; however, many synthetic chemicals are virtually indistinguishable from their biologically produced counterparts; so, distinguishing between chemicals on that basis would also be a very crude form of regulation. Finally, it is not simply the synthetic chemicals which pose hazards to health; many of the most toxic chemicals known to man are produced biologically (*e.g.* ricin from the castor oil plant and the botulinum toxin are both "biocides", toxic at doses much less than 10 microgrammes per kilogram), and some of the most potent carcinogenic substances in the food chain are naturally generated (*e.g.* aflatoxins).

Therefore the regulation of chemicals must be focused on the investigation of the specific effects that each chemical has on human and and/or environmental health. There is no simple categorical means by which safe and unsafe chemicals can be distinguished. Given the wrong conditions, any chemical can be unsafe. One can drown by inhaling water, but water is an essential for life, i.e. the unsafety is determined by the amount and the way in which the water enters the body. Effective safety regulation therefore requires the unsafe conditions to be established and for the chemical to be regulated so that these conditions are not met under defined circumstances of use or disposal of that chemical.

From this it is apparent that regulating chemicals by identity, biological properties and conditions of use in effect regulates also the mass production of chemicals. In the modern age of chemical production, it is now possible to alter the manner and quantities of the various organic chemicals produced at very short notice and in very dramatic fashion. Given this capacity, it is the more important to regulate so that the societally correct quantities of the various chemical characteristics are produced.

Once the core problem in the control of modern chemicals is perceived to be the attempted regulation of machines capable of producing a wide range of organic chemicals, the question then becomes how to regulate this "machine" for the production of the "proper balance" of chemical substances, whereby the maximum public benefit is achieved with only the minimum (i.e. acceptable) risk to the well-being of the public and the environment.

Every chemical has attendant benefits and costs implied in its use. It is the balancing of these which should be the primary object of the regulator. In the next section, the short-term (or acute) costs of synthetic chemicals are surveyed; then the longer term (or chronic) costs of chemicals are discussed; and finally, it is shown that a proper balance for the use of any given chemical exists.

Therefore this section was intended to demonstrate that the appropriate paradigm for the regulator of chemical inputs is a quantitative, rather than a qualitative, one. Although important qualitative decisions must remain (*e.g.* there may be some substances for which the non-harmful dosages are so minute, that quantitative regulation is rendered infeasible), the important point remains that the task is to regulate the correct quantities of given chemicals in society, not to separate the "safe" from the "unsafe". The latter task is a false target, since there is no such thing as a chemical that is "safe" at all levels of dosage.

3. Acute Toxicity of Synthetic Organic Chemicals

Acute toxicity relates to the size of the single dose (usually standardised on body weight) which has immediate harmful health effects. It is to be distinguished primarily from longer term, or chronic, health effects which arise from repeated exposure, often over an extended time course.

Most ex ante testing of chemicals, i.e. testing prior to the general release of the substance, is accomplished by means of animal experimentation.[6] Acute toxicity can be measured by a simple experiment which involves administering a range of oral doses of the chemical to sample groups of animals (usually mice or rats), and then observing how many in each group die within 24 hours.[7] A basic parameter used as a measure of the toxicity of

[6]The costs of using animals for this purpose are not considered in this paper, although it is evident that this is a very real cost to some parts of society and "humaneness" should be factored into the cost/benefit calculus in a proper "environmental costs" study.

[7]It should be noted that there are a range of acute toxicity tests used in toxicology, other than those which test only for mortality. The survey of chemical toxicology in this section is not intended to be exhaustive, only descriptive.

a substance is its LD50, which is the dosage which is lethal to 50% of the sample population. This dose is expressed in terms of grams of active ingredient per kilogram of body weight of the species tested.

Box 1. The Acute Toxicity of Synthetic Agrochemicals

Acute oral toxicity is the immediate harmfulness to health caused by the ingestion of a substance. The LD50 of a substance is the amount of the substance that is necessary to kill 50% of a sample group of animals within 24 hours. Experimental evidence indicates that there is a wide range of toxicity across both synthetically and naturally produced chemicals; that is, there is no clear dichotomy between "safe" and "unsafe" chemicals. This table surveys the acute toxicity of various "pesticides", which are chemicals that are constructed to be highly toxic to "target" species, and less toxic to others.

Supertoxic chemicals (i.e. rat oral LD50<10 milligrams/kilogram)		
Insecticides	Herbicides	Known Comparison
one organophosphate (endrin) several organophosphates (*e.g.* parathion) two carbamates (*e.g.* aldicarb)	no herbicides	deadly nightshade (atropine)
Higher toxicity (i.e. LD50<50 milligrams/ kilogram)		
Insecticides	Herbicides	Known Comparison
many organochlorines (aldrin, dieldrin, PCP) most organophosphates (dozen commercial substances) one carbamate	no herbicides	nicotine, cyanide and vitamin D
Moderate toxicity (i.e. LD50<500 mg/kg)		
Insecticides	Herbicides	Known Comparison
most organochlorines (DDT, lindane, chlordane) many organophosphates (diazinon) two carbamates most pyrethroids	most herbicides (2, 4-D, 2, 4, 5-T, paraquat)	barbiturates
Slight toxicity (i.e. LD50<5g/kg)		
Insecticides	Herbicides	Known Comparison
no organochlorines some organophosphates some carbamates/pyrethoids	most herbicides (simazine, atrazine, alachlor, diuron)	aspirin
Low toxicity (i.e. LD50>5g/kg)		
		Table Salt at 30 g/kg

Source: Adapted from [7] table 3.1; and [5] table 4.2.

There are several basic points illustrated in Box 1.

First, chemicals range across a spectrum of toxicity. There is no such thing as a "safe" chemical; there is only a "safe dose" level with regard to every chemical which should not be exceeded. Even salt, the most commonplace of compounds, is toxic if the safe dose level is exceeded. This is not meant to imply that all chemicals should be treated equally. There is a wide range of toxicity, and some commonly used agrochemicals (*e.g.* parathion and endrin) are more than three thousand times as toxic as the low toxicity substances, such as salt.

Secondly, agricultural pesticides are selected for their highly toxic characteristics to specific groups of animals/plants, so it should not be surprising that some of them fall into the higher groups on the spectrum of toxicity. However, species sensitivity can vary considerably with respect to any given pesticide.
Consider, for example, the evolution of the various forms of insecticides used in agriculture. The organochlorines were the first "modern" group synthesised for this purpose, and they were selected partially for their characteristic of "persistence", that is, due to their slow rate of decay, the insecticidal benefits of a dose of these pesticides continued without the need for repeated applications. In addition, as with all well constructed pesticides, they were targeted for high acute toxicity in insects, with a lesser effect on fish, and very low toxicity in humans. This made them ideal for controlling insect vectors of disease, such as mosquitoes and lice.

The persistence of these chemicals, however, has lead to their widespread distribution and thus to serious accumulation problems within the food chain. Low concentrations of these insecticides remained in the environment for long periods, allowing the higher animals (birds, mammals) progressively to accumulate these substances even when far from the area of application. The accumulation of these substances lead to population reductions in some animals (*e.g.* birds of prey) and even to annually increasing concentrations in the depot fat of humans.

Because of these problems, the organochlorines were banned in the 1960s by many western countries, and removed from the market in most others.[8] This shift was assisted by the development of the organophosphate chemicals, which were designed to be more toxic in order to compensate for their reduced persistence. Thus the result was that most of the agrochemicals used in the 1970s were in fact applied more frequently, and exhibited greater immediate toxicity than did those used in the 1960s. Subsequently, other types of insecticide with low human toxicities closer to the organochlorines have been

[8]Many organochlorines (*e.g.* DDT) are still used in the less developed countries, leading to the paradox that chemicals banned for use in one country are nevertheless produced there for use in another. This is, however, another example of the necessity to weigh the relative costs and benefits of any chemical usage. In the tropical countries, it was estimated by WHO that DDT had saved about 10 million lives in a campaign to eradicate malaria. Obviously, this benefit does not obtain in most of the more northern countries, and hence there is little need to incur the particular costliness of organochlorines for this reason. It is interesting to note that many of these "banned" chemicals are retained in stock in the developed countries by health departments in the event of a recurrence of disease for which they are particularly useful, *e.g.* typhus.

introduced (carbamates, pyrethroids); even so, the overall toxic load of insecticides is still increasing, because the average application rate of pesticide used in Europe is constant, while the area so treated is increasing [3].

In summary, there is no easy dichotomy between those chemicals which are safe or unsafe to use. All chemicals need to be regulated so that they are used only in proportion to their costs and their benefits. This applies equally to biologically and to synthetically produced chemicals. Both sources produce chemicals which range equally widely in their toxicity. It would be convenient if it were otherwise, but it is not. There can be no substitute for a detailed investigation of the specific health properties of each and every chemical substance, whether natural or synthetic.

4. Chronic Toxicity

Probably the greatest concerns about chemicals lie in their potential for long term impact, rather than in the prospect of an immediate poisoning. In particular, the risk of cancer from synthetic chemicals is of widespread concern [8]. Coinciding with the development of the synthetic chemical industry there has been a signficant alteration in the aggregate mortality rates. In particular, it is now far more common to have personal knowledge of cancer deaths than was previously the case, for this is now the primary cause of death in mid-life.

Box 2. Aggregate Mortality Statistics - Years Lost and Rates

Malignancies now constitute the primary cause of death "in the prime of life". More years of life are lost to accidents, but this is because this category applies almost exclusively to the very young. In addition, more lives are lost to heart failure, but this is primarily the fate of the very old. In mid-life, malignancies cause about twice as many deaths as all other causes combined.

Cause of Death	Annual Rate/1000	Years of Life Lost
Heart failure	293.5	1,566,000
Malignancies	192.3	1,809,000
Strokes	63.4	256,000
Accidents	42.2	2,367,000
Other lung diseases	22.9	114,000
Suicides/homicides	22.0	1,314,000

Source: Morbidity and Mortality Weekly Report (1985), Annual Summary 1984.

The coincidence of the chemical age with the increased mortality from cancer has, justifiably, raised the inference that there is a cause-effect relationship between them. However, it is in fact the case that the actual rates of all cancers (other than lung cancer) have remained constant or declined since 1930. The large shifts within mortality statistics have occurred not because of increased rates of cancer deaths, but primarily because of decreased rates of deaths from infectious disease.

Whereas the event of death was once something which might occur at any time throughout life by reason of such diseases, the availability of antibiotics (*i.e.* the application of manufactured chemicals which are toxic to human pathogens) has now shifted deaths from disease to middle age or later. This means that a greater proportion of any year's births are still alive in middle age and, even though the proportion dying from cancer in each year of life has not altered at all, a larger number of people will in fact die from cancer. The near elimination of deaths from infectious disease, with the consequent increase in the number of deaths from cancer later in life, has therefore reasonably resulted in the perception of an increase in the relative risk of cancer, and the consequent assumption that this is due to increased exposure to the products and by-products of modern industrial society, such as synthetic chemicals.

The actual risk of contracting cancer (at any given age) has nevertheless remained roughly constant over the past half-century. Between 1930 and 1980 the incidence of cancer in the U.S. population remained at around 300 per 100,000 each year.[9] Given this relative stability, the perceptibility of cancer deaths has increased dramatically against a background of otherwise declining causes of death in mid-life. Although the absolute level of risk has not increased during this time, the decline of other causes has made it appear relatively more risky. Therefore, the coincidence of the chemical age with increased awareness of the risk of cancer is more likely to be a beneficial consequence of pharmaceutical chemicals (*i.e.* antibiotics) than a consequence of its increasing riskiness.

Box 3. Best Estimates of the Causes of Cancer (1980)

The accumulated evidence on the sources of cancer risk indicate that the most easily detected sources are tobacco and diet. Most other risks are of an order of magnitude lower.

	Per cent of all Cancer Deaths	
Source of Risk	Best Estimate	Range of Acceptable Estimate
Tobacco	30	25-40
Alcohol	3	2-4
Diet	35	10-70
Food Additives	1	0.5-2
Sexual Behaviour	7	1-13
Occupation	4	2-8
Pollution	2	1-5
Medicines	1	0.5-3
Industrial Products	1	1-2
Geophysical Factors	3	2-4
Infection	10?	1-?
Unknown	?	

Source: Doll, R. and Peto, R. [1981], The Causes of Cancer: Quantitative Estimates of Avoidable Risks of Cancer in the United States Today", JNCI, v. 66 (6), p. 1191.

[9]This is not to say that the various forms of cancer have not exhibited differing mortality rates. In fact, the mortality rates from all cancers but lung have been declining in females since 1930, and constant in males. The incidence and mortality of lung cancer has increased almost fivefold for males between 1930 and 1980; its incidence in females has about doubled since 1950. Most of this change in trend is explained by tobacco consumption.

The best estimate of the causes of cancer in western society indicate that the most easily avoidable risk involves tobacco. Most other risks, with the exception of diet, are of an order of magnitude lower.

Again, this does not imply that modern synthetic chemicals are without their own chronic costliness, it is just that the first fifty years' record of use has not demonstrated a marked shift in such costliness. A recent study by the U.S. Environmental Protection Agency has listed 53 chemical compounds which are possible tumour-causing agents, including 4 pesticides listed as "probable" carcinogens and 15 others listed as "possible" carcinogens [9]. Although these are fairly small numbers as compared with the 64,000 commercial chemicals placed on the register since 1975, they do represent pervasive risks in our food. About half of the herbicides in use and over three quarters of the fungicides contain some of these listed compounds. However, the highest level of carcinogenic risk from any one of these substances was about 1 in 10,000 at the doses likely to be ingested, as compared with a 1 in 4 risk of contracting cancer from some cause in a lifetime [7].

5. The "Proper Balance" of Chemicals

It is important, therefore, to manage the intake of chemicals so as to procure a proper balance between them. For example, some of the most toxic chemicals are inorganic, and include the metals such as zinc and copper. Consumption of these metals in excessive amounts would soon be fatal. It is nevertheless essential to ingest very small quantities of these chemicals daily for maintenance of normal body functions; zinc and copper (and other metals) are essential for the continued operation of human enzyme systems. What is clear is that the proper dosage of these chemicals, although minuscule, is certainly not zero.

This is also true for the avoidance of chronic health risks. Food toxicologists indicate this fact in their study of impact of "dietary factors", which are listed as an important cause of cancer in Box 3. For example, butylated hydroxyanisole (BHA) has been shown to give protection against the deleterious effects of some known carcinogens when administered in low doses, whereas it itself initiates cancer in some rats when dosed at relatively high levels for prolonged periods [10]. The point is that while such a substance can never be termed either safe or unsafe it does provide an interesting example of a food additive in whose use careful consideration has been given to risk benefit analysis.

Sections 3, 4 and 5 have described the health-effect approach to the appropriate objective regarding the use of chemicals in society. The studies of chemicals, both synthetic and natural, indicate that the correct objective is the determination of the tolerable exposure to any given substance, weighing the costs attending further exposure against the benefits of their use. That is, the appropriate form of enquiry must be cost/benefit analysis, not just because of the important benefits which modern synthetic chemicals yield,[10] but primarily because there is no means of distinguishing between those chemicals which are

[10]The benefits from agricultural chemical usage are very real. Most estimates indicate that about 30% of crop production would be lost to pests without the application of pesticides.

"costly" (or "unsafe") from those which are not, without reference to dosage. The

Box 4. The "Proper Balance" of Chemicals

Studies demonstrate that diet is an important source of chronic health risks, but this is indicative of the importance of both its composition and its internal balance of chemicals. Some can be anti-carcinogenic at lower levels of use and pro-carcinogenic at higher levels. In general, a "proper balance" in diet demonstrates the lowest health risk. For example, it has been found in Japan that omnivorous persons face a lower risk of colon cancer than do either pure vegetarians or pure carnivores.

Dietary Pattern	Colon Cancer Risk (per 100,000)
Meat but not vegetables daily	18.43
Vegetables but not meat daily	13.67
Meat and vegetables daily	3.87

Source: Hirayama, T., [1985] "Diet and Cancer: Feasibility and Importance of Prospective Cohort Study" in Joosens, J., M. Hill and J. Geboers, (eds.), *Diet and Human Carcinogenesis*, Elsevier: New York.

regulation of the mass production of the organic chemicals therefore requires the careful investigation of the costs and benefits of the output and use of each chemical.

6. Systemic Defects in the Regulation of these Risks

This section is intended to demonstrate how the market works to compare costs and benefits even in the absence of an explicit regulatory framework. For many commodities (even dangerous ones), the market alone is capable of determining the "correct" amount of the hazardous activity.

This, however, is probably not the case for synthetic chemical introductions. The difference lies in what is known as the economic concept of "externality". When an agent in the market is responsible for undertaking a cost/benefit analysis for society in determining the correct quantity of a given input to use, it is important that this agent bears the full costs of its decision. Otherwise, if some of the decision's costs are borne by others (and thus "externalised"), the market's outcome will be incorrect. In this case, governmental regulation of a specific character is indicated.

What follows is a discussion of the nature of those market failures that make governmental regulation of synthetic chemicals necessary, and of the nature of the regulation that is indicated. This is set in a hypothetical world where no regulation is in place initially, in order to demonstrate the sort of measures that would be required in these circumstances. Later sections compare these "model forms" of regulation with those presently existing in the EU.

As mentioned above, there is an in-built market mechanism which attempts to integrate society's costs and benefits in the use of chemicals in agricultural production - this is the farmer. Agricultural chemicals are just one of several production factors (e.g.

labour, machinery, land, seed) which the farmer combines in order to produce his output. In determining the portion of chemicals to use in generating any given level of output, the farmer will weigh the costs of additional chemical usage against the prospective benefits from its use.

The costs which the farmer will consider are derived from the price to him of an additional application of chemicals. To a degree, this price represents society's costs as well, because the farmer will have to pay enough for the chemicals to divert their use to agricultural purposes: For example any petrochemical product can be used in a variety of different fashions (plastics, fuels, chemicals). The price that each of these industries pays must be sufficient to bid the petrochemical away from the others. This is the "opportunity cost" of the chemical.

Box 5. Excessive Chemical Use in Agriculture (Without Regulations)

The farmer makes the actual decision on the balance of chemicals to apply in producing agricultural goods. He will balance, to a degree, the costs and benefits to society of additional chemical inputs in making this calculation (society's costs and benefits are demonstrated by the price of the chemical [PMC in graph] and the price of the additional product [MRP] respectively). However, some of society's costs - the health risks of chemicals - will not be reflected in the price which the farmer pays for chemicals; the actual costs of chemical applications should be above PMC, i.e. SMC. The farmer's cost/benefit analysis is insufficient; this would result in systematic over applications of agricultural chemicals. That is, without government regulation, the perhaps hazardous amount (Q) rather than the societally optimal amount (Q*) of synthetic chemicals would be applied.

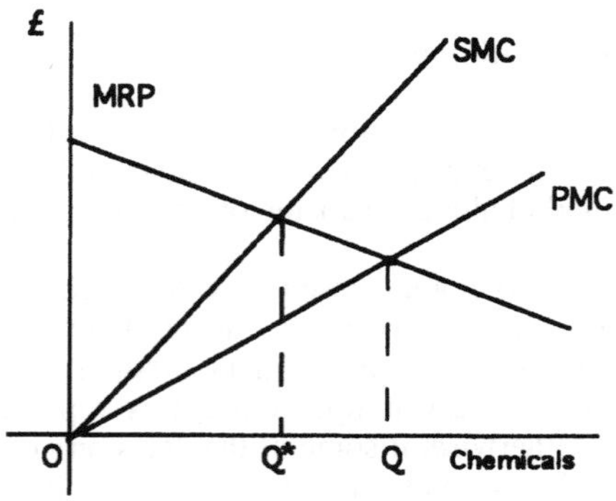

The benefits which the farmer will consider derive from the additional revenue which this application of chemicals will bring to him; this will be the product of the additional yield generated by the additional chemicals and the price of the produce. Again, this is equivalent to society's objective; the price of the produce indicates society's aggregate willingness to pay for more of that particular commodity versus all others.
Therefore, the cost/benefit calculus considered by the profit maximising farmer in attempting to determine the correct proportions of various chemicals is, to an extent, identical to society's cost/benefit calculus in determining the "proper balance" of

chemicals.[11] However, the farmer's calculus will also deviate from that of society's in certain critical ways.

The systemic defect which prevents the farmer's cost/benefit analysis from being a reasonable approximation of society's is the absence of the costs of potential health risks to persons other than himself (and to other environmental aspects such as wildlife) from the farmer's calculations. The farmer only considers the costs of depriving the plastics and fuels industries of the petrochemical resource; he does not consider whether there might be potential acute or chronic health costs associated with his application of chemical inputs. This cost is never fully reflected in the price which he has to pay for the chemical input; that part is external to his decision-making framework. This systemic defect could result in an inappropriate quantity of chemicals being applied in the production of agricultural products especially if increasing application led directly and consistently to increased profits.

Since there are millions of farmers in Europe making these calculations, the role of the regulator should be to apply a correction to this systemic defect in the farmer's decision-making framework, rather than attempting to replace them entirely with another cost/benefit analysis machinery.[12]

The environmental economist would advise that, in order to acquire the information necessary to successfully alter the decision making of the individual farmer, the regulator's object should be to:

(1) determine the level of health costs corresponding to various aggregate levels of intake of any given agricultural chemical (i.e. determine a continuous dose-response relationship and value the costliness of the response);

(2) determine the level of intake corresponding to any given aggregate level of chemical application (*i.e.* determine the amount of residues taken - and perhaps accumulated - by humans from various media, including food, given the chemical application); and

(3) determine the rate of application at the level of the individual farmer which corresponds to any given level of the aggregate application (*i.e.* determining the relationships between individual and industry behaviour).

The completion of these three tasks would determine the difference between social and private costliness in the farmer's decision-making framework (*i.e.* the difference

---

[11]In fact, it would be precisely identical if consumers were accurately informed of the risk level associated with consumption of the products which have chemical inputs. Then they would discount their willingness to pay for the risk implied, and the prices facing the farmer would fall for increasing chemical inputs [11]. This is the rationale for the creation of "organic produce" markets.

[12]The alternative would be to "nationalise" this decision making function; that is, the government could tell farmers precisely what proportions of chemicals to use. There are two problems with this. One, so long as the chemicals may be acquired on the market, there would be the difficulty of monitoring each of these millions of farmers for compliance with instructions. Regulators can tell you of the costliness of this exercise [12]. Secondly, there is little left to the job of being a "farmer", once the responsibility for determining factor levels is removed. Regulating society through further centralisation removes many of the incentives for individuals to be innovative or even productive.

between SMC and PMC in Box 5, above). The assessment of a chemical use charge (t) corresponding to this difference would internalise the health costs of agricultural chemical usage within the farmer's cost/benefit analysis. The amount of agricultural chemicals applied would then include the consideration of this costliness; the systemic defect in the farmer's cost/benefit analysis would be corrected. Equivalently, the regulator could determine the aggregate amount of a given chemical that would be applied (Q* in Box 6.5) and allocate rights of application to the various farmers. The two methods of proceeding are theoretically equivalent under certainty; both result in Q*, the socially desired aggregate amount of the chemical, rather than Q, the unregulated result.

Therefore, under certainty, the regulator's task is easy and straightforward. That is, it involves the setting of the correct societal application rates of various chemicals by means of looking for examples of uninternalised costliness. The existence of these external costs then indicates the need for societal restriction or taxes, but not necessarily prohibitions, on the use of the chemical. This is how, in theory, the regulator's task is formulated in conjunction with the economic workings of the market.

7. Regulating Uncertainty Regarding Chemical Usage

The problem with the above scenario is that it does not describe the reality of chemical regulation very well. The very essence of synthetic chemical regulation lies in the acceptance of the uncertainties that exist. That is, there can be little information *ab initio* with regard to the wide range of potential effects of substances that have never existed before.[13] The implications for efficient regulation of working in the presence of substantial uncertainties are the subject of the remainder of this chapter.

Studies in the fields of environmental toxicology and environmental economics indicate that any one of the three regulatory tasks listed above is very difficult to accomplish with a high degree of accuracy, even for very clearcut health risks. One of the most difficult tasks is the valuation of life-endangering activities. This is a very controversial concept, and it yields widely ranging estimates of the costs of such activities to society. For example, occupational safety studies yield implicit life risk values in the U.K. ranging all the way from $15,000 in the agricultural industry to $10,500,000 in the pharmaceutical industry, although the consensus figure lies at about $500,000).

Nevertheless, the single most difficult task to be accomplished in the regulation of chemicals in agricultural usage is the regulation of uncertainty. Uncertainty pervades the entire health risk assessment process regarding chemicals, because *ex ante* testing can only reduce the range of possibilities, not predict all outcomes with certainty. In addition, there are some risks from chemicals which are not directly amenable to *ex ante* (i.e. prior to human exposure) assessment.

The regulation of these two forms of uncertainty should be the focus of chemical regulation. This is because the modern chemical age is an age of mass production,

[13]This is not literally correct, as much information can be acquired by analogy between similar but distinct chemical substances. However, this does not obviate the fact that new permutations create new uncertainties that can not be eliminated by analogy, but only by testing.

implying abrupt rather than gradual shifts in technology. It can be expected that a cost-effective chemical will be adopted universally across the agricultural industry almost immediately upon its commercial release; otherwise, (due to the competitive nature of the industry) the crop (in which this particular chemical is a necessary input) can not be growth cost effectively. Therefore, there is no small scale opportunity to watch for unforeseen effects of a chemical, as there would be if the chemical were to be gradually introduced into the environment. This raises the issue of *ex ante* testing to pre-eminent importance; *ex post* information (*i.e.* epidemiological studies) may be acquired only at the expense of significant health costs and even environmental irreversibility. Therefore, the regulation of the testing procedure is of the first importance in this context.

8. Reducible Uncertainty and *ex ante* Testing

Reducible uncertainty corresponds to the range of possibilities which remains after a given test result. There are two particular sources of such uncertainty in laboratory testing. First, there is the uncertainty which remains because of the small size of the test sample (compared to the ultimately exposed population) and the bias which this smallness has on expected effects (because the sample group may not be typical of the population from which it was selected, or may be too small to reveal an effect of low prevalence). Second, there is the uncertainty which remains because of the specific conditions which obtain in the test (*i.e.* the laboratory conditions) and the inability to generalise from these conditions to the range of conditions which exist in the real world. Both of these ranges of uncertainty are reduced each time that a test is repeated (on a different subject or under different conditions), but uncertainty cannot be reduced to zero unless the tests are repeated an infinite number of times; therefore, some reducible uncertainty must always remain. The precise amount depends on the number and/or the types of tests which are conducted. Again, this points to the necessity of balancing the costs of further reducing uncertainty against the benefits from further reduction.

The uncertainty resulting from the number of animals in the sample study is mathematically derived. In essence, given the natural variety of individual reactions to a given dose of a given substance, it is to be anticipated that a small sample of exposed subjects might have reactions atypical of those which would actually occur throughout the entire population; this is because the small group might by chance be made up of a number of individuals who are unusually susceptible or unusually tolerant of the substance. As the sample group becomes larger, so the probability that reactions to the chemical will be more validly predictive of the population response becomes greater.

Mathematically, the "power" of statistical testing in reducing uncertainty rises with the square root of the number of animals in the group. This is an example of what economists know as the "law of diminishing marginal returns". Although further testing will always reduce uncertainty further, it does so in marginally decreasing amounts. Therefore, even if the cost per test is constant, there will come a point where the cost of further reducing uncertainty through this means will exceed the benefits of further reduction in these forms of uncertainty.

There is another form of uncertainty which is reducible through further *ex ante* testing; this form derives from the design of the test rather than the number in the sample.

That is, any given test procedure will involve a particular combination of factors presented to the sample group in the laboratory, and the hypothesis to be tested is whether this combination of factors (including the dosage of chemical) will result in a higher incidence of disease. If the sample group does exhibit a higher incidence, then the hypothesis of

Box 6. Number of Animals Required to Show Statistical Significance of Carcinogenic Potential of Chemical Substance

The sample group of animals must be increased in order to reduce the uncertainty that their reaction to a particular chemical substance is not atypical of the whole population. This uncertainty is a function of several factors: 1) the "background rate" of tumours (*i.e.* the rate at which tumours develop in the population over time irrespective of the specific exposures); 2) the desired "level of confidence" (*i.e.* the degree to which population variability is to be ruled out as the causal factor - this can never be 100%, and is usually selected to be 90% or 95% confidence); 3) the minuteness of the risk to be detected by the test (*i.e.* the capacity of the test to detect risks only marginally above "background risk" - for example 5% or 10% above). An example of the number of test animals required to meet these criteria while detecting a carcinogenic effect 80% of the time is given below. This table demonstrates that, to have 95% confidence in a test being 80% accurate, it is necessary to test 155 animals in order to detect a risk which is 10% above (a 10%) background risk or 540 in order to detect a risk which is only 5% above that background risk.

	Number of Animals Required vs Minuteness of Risk	
Background Rate (%)	(5% above)	(10% above)
1	143	55
5	330	105
10	540	155

Source: Lave, L. (ed.) [1982], Quantitative Risk Assessment in Regulation, Brookings Institution: Washington.

positive impact cannot be rejected; this means that, at the given level of confidence, the combination of factors might be carcinogenic. On the other hand, the failure to produce the statistically observable result does not imply the acceptance of the hypothesis that there were no disease-causing agents in the experiment. All that is implied is that this particular combination of factors (of laboratory air, feeds, light, stress, activity, etc.) did not give rise to a positive result in this particular test.

Therefore, laboratory tests give rise to "spotty" information with regard to the absence of health effects. This is the problem of the "false negative"; a failure to produce a significant health effect in a given experiment cannot prove safety - it only proves that within the limits of the experiment no risk was observed. A good model for this problem is to compare the experimental process to looking for a yellow tennis ball in a pail (full of white balls). If a yellow tennis ball is observed in the top layer of the pail, then the proposition that it is in the pail cannot be rejected. However, if all that is observed are white tennis balls, this does not mean that no yellow balls are in the pail, only that the experimental procedure used did not observe them [13].

A larger number of experimental procedures may be successful at discovering an earlier unobserved health effect (at a particular dosage), and in any event successive procedures will narrow the uncertainty about its existence. Observing the top layer of tennis balls in the pail gives one level of confidence, while combining this with a method which also observes the bottom layer (*e.g.* by drilling a hole in the bottom of the pail) will narrow the range of uncertainty even further. Therefore, the performance of a sequence of tests is necessary in order to narrow the range of uncertainty concerning the lack of harmful effects of a chemical at a given dosage.

The number of combinations of factors which might influence the disease potentiality of a given substance is nevertheless infinite, which means that uncertainty from this source can never be entirely eliminated either. At a given dosage, it would be necessary to vary all of the various combined inputs (through air, diet, light, activity, etc.) across the whole of their possible ranges. This means that the proposition of "no discernible health effects" cannot be accepted with certainty until all of these combinations have been attempted; this implies that the aggregate cost of weighing the benefits of further testing (in reducing the range of uncertainty) against the costs of yet another test.

An excellent demonstration of the means of implementing cost-effective *ex ante* testing is given in a recent study of carcinogenic testing. This study proceeded by means of accepting as accurate the previous experimental results on 38 chemical substances (which had been thoroughly vetted in animal experiments) with regard to discernible carcinogenicity at a given dosage. It then subjected these substances to a series of 64 *in vitro* tests (*i.e.* not animal tests) in order to assess which of these tests or which combination of tests would most often reach the same conclusion as the previous animal testing. Given the capacity of each of the 64 testing methodologies to predict whether the chemical substance would be carcinogenic in animal studies or not, an econometric analysis then determined what degree of accuracy was obtained through the use of any one test or any combination of tests. By comparing and contrasting these results, it is possible to construct an estimation of the cost of procuring less uncertainty through further testing.

Box 7. The Marginal Cost of Reducing Uncertainty

Sixty four different tests were compared for their capacity to predict the carcinogenicity of a given chemical substance (at a given dosage level). It was found that three tests were able to achieve the greatest degree of cost-effective accuracy in isolation: test numbers 19 (65% @ $400), 13 (17% @ $1200) and 21 (79% @ $800). In combinations, tests 13 and 54 (80% @ $6000) and 13 and 21 and 54 (89% @ $8000) performed most cost effectively. From this set of data, it is possible to determine a schedule for the cost of removing uncertainty.

Certainty Level	Increase in Certainty	Marginal Cost
65%	65%	$400
79%	14%	$400
89%	10%	$7200

Source: Lave, L. and Omenn, G. [1986], "Cost-effectiveness of short-term tests for carcinogenicity", Nature, v. 324 (6092), p. 29.

In order to provide the correct amount of *ex ante* testing, a regulator must compare the marginal costs of further uncertainty reductions (here, $7200 for the last 10%

reduction) against the benefits received from such a reduction. The benefits to be received from a reduction of uncertainty derive from the avoidance of both "false negatives" (*i.e.* the possibility of not detecting a dosage that has harmful health effects in a given test). The benefits of further testing will be dependent upon: 1) the proportion of substances which are health risks (*i.e.* the extent to which the health risk which is being investigated exists); 2) the costliness of a "false negative"; 3) the costliness of a "false positive".

The first factor is important because of the limits to the accuracy of statistical techniques in detecting "rare events". Therefore, it is very difficult to expect great benefits from a testing procedure capable of achieving 95% accuracy if it is being used to assess a risk which is much lower in magnitude, say 1%. The testing procedure will be of much greater usefulness if the risk were present in 5% or 10% of the cases. Essentially, this points to the importance of using low-cost methods for doing the initial screening of substances to be tested, so that there is greater confidence that harmful effects are then detected by reason of their existence (rather than by chance).[14]

The other benefits of further testing derive from its relative accuracy; and the costliness of false negatives and false positives. "Accuracy" can be defined in different ways, by weighting the relative importance of a false negative as compared with a false positive. The above figures give the accuracy of various techniques weighting an inaccurate negative finding the same as an inaccurate positive finding. This is not only unnecessary, but probably undesirable, given the differing costliness of the two results. A false positive results in the removal of the product from the market (at least at that dosage level) whereas a false negative means that unpredicted adverse health effects will occur.

Testing can have very different capacities with regard to predicting negatives and positives; for example, one test in the study above scored 26/26 on non-carcinogens but only 5/12 for carcinogens. That is, it had an in-built bias toward finding non-carcinogenicity. Therefore, it is fundamentally important that a weight be given to the value of both adverse results, so that a poor test is not mistaken for a good one.

Therefore, reducible uncertainty can be dealt with through the cost/benefit analysis of further *ex ante* testing. The correct amount of *ex ante* testing corresponds to that amount which achieves the proper balance of costs and benefits. It balances the costs of conducting one more test against the benefits deriving from the reduction of uncertainty from that test.

Therefore, the economics of optimal testing gives direction to the regulator with regard to the means of regulating uncertainty, as well as known risk. In the context of agricultural chemicals, it is fundamentally important that the correct amount of testing of any chemical substance occurs prior to its mass production. Currently, every chemical proceeds through about 6 years of testing at a cost of about $1 million prior to its release in

[14] This constraint leads to the "low probability event paradox'. For example, suppose that 60 of 1,060 chemicals are carcinogenic at a given dose. Given a test which determines "positives" correctly with 95% accuracy and "negatives" correctly with 94% accuracy, what is the probability of a "false positive"? A substance from this group which is determined to be carcinogenic according to this test has in fact only a 9% probability of being a carcinogen. This result derives from the relative inaccuracy of the testing procedure as compared to the proportion of actual carcinogens being sought after in the population (about 5%) [13]

most developed countries; however, there are many previously marketed chemicals (about 64,000) which have never been tested to modern standards. In addition, there is little analysis of whether this period and extent of testing is efficient. The procedures have been developed more in response to the seriously harmful effects of a few chemicals, rather than in regard to the average risk posed by a new chemical.

Box 8. The Regulation of Reducible Uncertainty

Reducible uncertainty should be regulated by means of determining the correct amount of *ex ante* testing. This is done by comparing the marginal costs of further reductions in uncertainty (MC) with the marginal benefits of further reductions (MB).

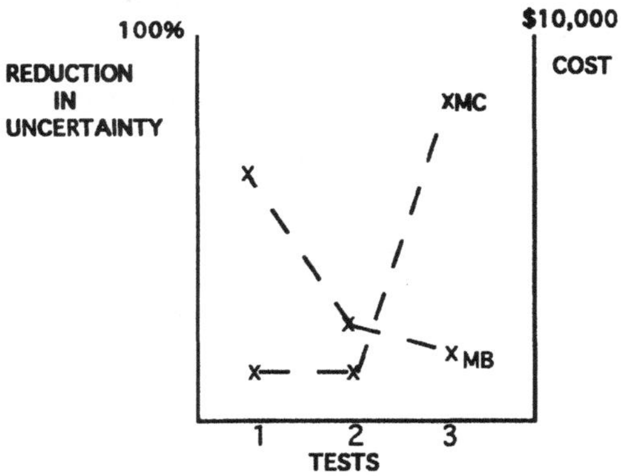

It should also be noted in regard to ex ante testing that, to some extent, this testing activity may be relaxed if our methods for reducing chemical riskiness are used instead. To some degree, the need to reduce uncertainty depends on the relationship between the doses that are shown to be harmful (object 1 or the regulator) and the dose likely to be experienced by the public in practice (object 2 of the regulator). The extent to which the first exceeds the second is a "margin of safety", and the greater this is, the less important are the associated uncertainties of object 1 and thus the amount of *ex ante* testing required to reduce it. Therefore, the better design of chemicals (for lower toxicity, lesser persistence, better targeting) is an alternative mechanism for reducing aggregate chemical costliness; when this is done very well, the need for testing is reduced although never eliminated.

9. Regulating Irreducible Uncertainties

There are some forms of uncertainty which are never reduced, irrespective of the amount of *ex ante* testing conducted; these are the irreducible sources of uncertainty. In essence, there are some effects which are not predictable under existing methodologies and will not be discovered until after human exposure occurs on a large scale. Largely, this manner of uncertainty derives from extrapolating to humans from findings obtained with other species.

There are two related forms of uncertainty. First, there is the uncertainty of extrapolating from short term exposure to longer term. Because of man's longer life span, it is not in one sense possible to perform animal studies testing for long-term exposure to low doses of chemicals. In order to achieve a lifetime dose accumulated from small daily doses would require a lifespan identical to humans, but there are few species of mammals that are able to match this. Thus, most testing occurs over much shorter absolute spans, with much greater daily doses.[15] These studies provide a rough basis for comparison with human long term chemical exposure because the test animals are nonetheless dosed for the greater part of their natural lifespans and the higher dose used increases the chances of detecting any adverse effect which the chemical may be capable of causing.

Box 9. The Sources of Irreducible Uncertainties

Ex ante testing can reduce uncertainty only to the extent that the health effects of substances on other species provides a reasonable guide to their effects on humans. This is not always a very clear cut matter. For example, the pesticide dieldrin is 25 times more lethal for rabbits than is the pesticide carbophenothion; however, in dogs the relationship is the reverse (i.e. carbophenothion is three times as lethal as dieldrin). It is helpful in such cases to select as the surrogate for man a species in which the chemical acts and is both qualitatively and quantitatively handled in ways close to those in man.

	LD50 Values mg/kg	
Species	Dieldrin	Carbophenothion
Rat	38-87	7-91
Mouse	38/74	106-218
Rabbit	45-50	1250
Dog	56-120	40
Chicken	48	316

Source: Stanley, P. and Bunyan, P., [1979], "Hazards to wintering geese and other wildlife from the use of dieldrin, chlorfenvinphos and carbophenothion on wheat seed treatments", *Proc. of Royal Soc. Lond.*, **B205**, 31.

Since science accepts that a negative can rarely if ever be proved, there is little to be done about reducing these forms of uncertainty through *ex ante* testing methods. This does not mean that there is no method available for cost-effectively reducing these uncertainties.

In particular, it is important to recognise that the registration of a chemical for general usage does not imply the end of the testing process. Instead, an equally important

[15] In fact, it is probably the case that few regulators have ever considered requiring an actual human lifetime animal study in order to examine for chronic effects; this has as much to do with the incentive system facing regulators and chemical companies as it does with animal lifespans. The assumption of this chapter is that, if the cost/benefit analysis required a longer term of study for chronic effects, then such a study would still be constrained by the lifespans of the species under consideration.

component of the testing process must take into account the costs from both "false negatives" and irreducible uncertainties. These residual risks impose costs on society with certainty, when the whole body of registered chemicals is considered.

In essence, the approval of a chemical by government authorities constitutes the completion of its *ex ante* testing procedure and the commencement of its *ex post* testing procedure. *Ex ante* testing cannot "prove" the safety of a chemical substance, so regulators must invest resources into devising and implementing correct *ex post* testing mechanisms. It is essential that both the pre- and post-registration phases with regard to a new synthetic chemical be equally important parts of the chemical testing procedure.

Of course, this has been important in the pharmaceutical industry, where the identity and amount of the drug received by a sensitive individual is usually well-documented. It is more difficult to achieve with agrochemical pesticides where exposure dose as a result of normal usage is more difficult to identify and the cause of an isolated harmful effect may be less easily substantiated. A more sophisticated form of monitoring would be required than presently exists in the pharmaceutical industry.

There can be no substitute for human-specific data for reducing the uncertainties that remain; however, the costliness of acquiring this data could be minimised (after a substance had successfully completed the *ex ante* testing hurdles) by phasing in the introduction of a substance over time and in a fashion which lends itself to isolating health impacts (*e.g.* limited local use for a number of years).

10. The Existing Pesticide Registration Scheme

National pesticide registration schemes require the manufacturer to provide a considerable amount of information on the efficiency, and on the mammalian and ecotoxicological properties of a new product. This entails a considerable expenditure during the four years (approximately) that it takes to develop the product. At regular intervals during this period, assessments are made by the manufacturer to determine whether the product will be acceptable to the national registration schemes; the development will be aborted if there are significant doubts in this respect, since failure to obtain registration will entail substantial loss of development phase investment.

Over the past 30 years, most countries have developed their own national registration schemes. Many of these schemes require that specific tests be carried out on the product. A pesticide manufacturer therefore has the obligation to carry out a number of different tests in order to register the product in different countries. Some of these tests may be simple duplications, e.g. 96 hours LC50s for different species of fish, where data from one fish may be sufficient to reduce this expenditure.

Even where the information requirements are similar, there can be different procedures used in the assessment of the potential hazards and risks associated with a product. For example, a review of the extent to which European countries used fish toxicity data in the assessment procedure revealed a wide divergence between zero and considerable emphasis. Also, the benefits of using the product are taken into account; if a benefit is peculiar to a single country, registration and approval of a product will be easier there than in another where there is no expectation of benefit.

For these reasons, product registration in one country does not necessarily assure registration in another, unless the former country's standards are equally stringent and the nature of the benefits is the same.

Until recently, the costs of product registration in the UK were borne by the regulatory authority, but now they are incorporated in a pesticide registration fee. Also, the costs of any additional tests required by the authority are borne by the industry; in the UK the need for such tests is agreed with the manufacturer. The risk of potentially expensive additional testing at this stage is one borne by the manufacturer.

Where there are some doubts about the efficacy of potential environmental effects of a pesticide, approval can be given for a limited usage, with further data being acquired before a decision to grant full approval is given. In all cases, approval is limited to the pesticide being used at a given rate on specific crops or target pests. Any extension of use requires further approval. Other restrictions and safety advice are given on the product containers.

So far this review covers the *ex ante* situation. There are three main components in the existing *ex post* process: monitoring, accidents, and the appearance of unacceptable effects.

Monitoring of foods and the environment (e.g. water, wildlife) is carried out to determine whether the pesticide concentrations have exceeded the acceptable limits set for them. The costs of this monitoring are not borne directly by the industry in the U.K., but by a fixed levy on the annual turnover for each product. To some extent this is a fair approach because a product with considerable sales is more likely to cause environmental contamination than is a little-used product. However, monitoring frequency is sometimes linked to the risk of damage, and under the present scheme all pesticides are treated equally in this respect. High volume, low risk pesticides (e.g. some herbicides) may incur a disproportionate financial burden under this scheme.

Most pesticide accidents lead to acute exposure of humans and/or the environment. Regulations are laid down for pesticide operator safety, the safe disposal of pesticide containers, and the containment of stored and spilt products. All accidents have to be reported to the appropriate regulatory authority; identification and assessment of damage is relatively easy and the penalties and compensation can be set by liability laws.

More difficult is the identification of insidious and perhaps progressive environmental damage which may become apparent only a considerable time after the pesticide has been approved for use. Such undesirable effects can be overcome by controls on the use of the product or by a complete ban. A particularly troublesome problem is one which is dectected only after a product has ended its commercial life, and its manufacturer is no longer in business.

Restrictions on pesticide use carry with them additional monitoring requirements. For example, about thirty years ago the use of dieldrin to control ectoparasites of sheep was banned. The alternative pesticides were not as persistent as dieldrin, nor were they as effective. However, dieldrin formulations for the treatment of woollen goods against moth attack remained available and these were used illegally by some farmers to boost the replacement products. The problem came to light when effluents from wool processors were found to contain dieldrin and such monitoring has to continue so long as this pesticide remains available.

A second example relates to a change in pesticide usage. Dichlorvos is an insecticide widely used to control ectoparasites on intensively reared chickens. The formulation used was subsequently found to be effective in controlling sea-lice on salmon in marine farms. Concern was expressed about the possible effects of such usage on native stocks of crustacea, e.g. crabs and lobsters. Approval was given for a limited use of this product, but the rapid expansion of the salmon farming industry led to its widespread use. Approval for this wider use depended on the outcome of a series of ecotoxicological tests and environmental surveys. These were expensive, and the manufacturer stated that because the use of the product for this purpose was comparatively small, it would withdraw it from this market. There were no suitable alternative products available. A ban on the use of dichlorvos could have led to the collapse of the salmon farming industry which had been developed in economically poor areas with considerable public financial support. There was no indication that the current usage of dichlorvos was having an adverse effect, although it could not be assumed that no effect existed. Clearly, the benefits of the pesticide use will weigh heavily in assessing the proper balance in such a situation. Proper apportionment of the additional research costs is a difficult problem.

The difficulty of establishing an *ex post* regulatory regime (and allocating its costliness) remains. For example, the use of tri-butyl tin (TBT) in boat anti-fouling paint was found to have an adverse effect on oyster farming in some estuaries, particularly where there was a high concentration of pleasure craft (e.g. in marinas). It took several years of intensive research to establish this dose-effect relationship beyond reasonable doubt. The link was strongly disputed by the paint manufacturers until they were in a position to place a more environmentally acceptable product on the market. Again, it is important to allocate efficiently the cost of identifying the cause of this environmental damage, and to return the costs to the polluters; sometimes the former object will be more costly than the latter. To some extent this example is similar to that of atrazine in ground water that forms the basis of this project.

This also raises the questions of the responsibility of the regulatory authority for the pesticides that receive approval and subsequently are shown to cause problems. To some extent the cost of carrying out further research may be re-allocated to the manufacturer if the approval for use is limited to, say, five years. Re-registration after this period may result in a request for the provision of additional information before approval is given.

One thing is clear. It is imperative that the science of ecotoxicology is developed so that predictions or the establishment of concentration-effect relationships can be made with the minimum of experimentation and with the maximum of accuracy. Also, environmental monitoring should be as economical and informative as possible. The need to achieve these goals cost-effectively should be seen in the context of the economic framework of pesticide use, and the need to achieve a proper balance of benefits and costs. We are a long way from achieving this ideal situation; in part this is due to the high public and political profile of pesticides which can lead to the adoption of a conservative approach concerning these chemicals. As the perceived need to be "on the safe side" increases, so the need for accurate scientific assessments decreases. To some extent, this demand for a large margin for safety explains the lack of scientific progress in this field.

11. Conclusion

Regulating chemicals in our society is not a task capable of simplistic solutions; neither "bans" nor "simple registrations" are capable of producing the correct balance of chemicals for maximum human benefit. The single most important feature in the regulation of chemicals is the construction of a process which is developed to cost-effectively reduce uncertainty, both *ex ante* and *ex post*. Only after this procedure is in place can the mass production of chemicals be set or reset at a proper level. The extent to which *ex post* (*i.e.* epidemiological) investigations are necessary and cost-efficient (in lieu of greater margins of safety, for example) for pesticides is an issue requiring further analysis.

References

1. OECD [1983], *Economic Aspects of International Chemical Controls*, OECD: Paris.
2. Davis, D. and Magee, G. (1979). Cancer and Industrial Chemical Production, *Science*, 306, 1356.
3. World Resources Institute [1990], *World Resources 1990-91*, Basic Books: New York.
4. Bull, D. [1982], *A Growing Problem: Pesticides and the Third World Poor*, Oxfam: Oxford.
5. Crone, H. [1986], *Chemicals and Society*, Cambridge University Press: Cambridge.
6. Sittig, M. [1980], *Pesticides Manufacturing and Toxic Control Encyclopaedia*, Noyes: New Jersey.
7. Conway and Pretty, [1991], *Unwelcome Harvest*, Earthscan, London
8. Brown, J. (ed.) [1989], *Environmental Threats: Perception, Analysis and Management*, Belhaven: London.
9. NRC [1987], *Regulating Pesticides in Food*. National Research Council: Washington, D.C.
10. Pariza, M. [1989], A Perspective on Diet and Cancer, in Taylor, S. and Scanlan, R. (eds.) [1989], *Food Toxicology: A Perspective on Relative Risks*, Marcel Dekker: New York.
11. Rosen, S. [1981], Valuing Health Risk, *American Economic Review*, v. 71(2), p. 241.
12. Andreasson, I. [1989], *Costs of Controls on Farmers' Use of Nitrogen*, Stockholm School of Economics: Stockholm.
13. Page, T. [1978], A Generic View of Toxic Chemicals and Similar Risks, *Environmental Law Quarterly*, v. 7, p. 207.

THE EC DIRECTIVE ON DRINKING WATER: INSTITUTIONAL ASPECTS

M.G. FAURE
Professor of International and Comparative Environmental Law
METRO, Institute for transnational legal research
University of Limburg
P.O. Box 616
6200 MD Maastricht
The Netherlands

Part 1
Economic aspects of environmental regulation: basic concepts

1.1. INTRODUCTION

In the economic analysis of law, considerable attention has been paid to the question of whether efficiency can be reached through a market solution or through regulatory interventions by the government. In particular, studies have investigated whether harmful activities should be prevented by using the deterrent function of tort rules or by setting specific standards of care in regulation. Steven Shavell, who is currently professor of law and economics at Harvard Law School, has studied the alternative use of liability rules and regulation to prevent accidents[1].

There are a number of important differences between safety regulation and liability rules in so far the intervention of the government is concerned. Indeed, if one uses liability rules, the standard of care is not set by the government, but either by the parties in an accident setting (with a strict liability rule) or by the courts (with a negligence rule). This is totally different from the case of safety regulation, where the government will determine *ex ante* the standard of care with which the parties in an accident setting have to comply. If parties do not follow the standard, public sanctions (such as administrative or criminal sanctions) will be imposed. An important difference between both systems as far as enforcement is concerned is that with regulations the sanctions always apply in case of

1. See Shavell, S., "Liability for harm versus regulation of safety", *Journal of Legal Studies*, 1984, 357-374; Shavell, S., "A model of the optimal use of liability and safety regulation", *Rand Journal of Economics*, 1984, 271-280 and Shavell, S., *Economic analysis of accident law*, Cambridge, Harvard University Press, 1987, p. 272-290.

L. Bergman and D.M. Pugh (eds.), Environmental Toxicology, Economics and Institutions, 39-87.

non-compliance, whether or not damage has occurred, whereas tort liability only occurs if an accident happened.

In this part of the paper we will give an overview of the criteria pro and contra regulation. These criteria can also be applied to the harm caused by environmental pollution. Thus the criteria will allow us to conclude whether environmental pollution can only be controlled by using the deterrent effect of tort rules or whether a regulatory intervention is necessary, e.g. by setting an emission standard *ex ante*.

We will first formulate the problem of environmental pollution in economic terms (Section 2), then look at the criteria for safety regulation (Section 3). We will then address the question of whether the existing environmental regulation can always be explained on efficiency grounds (Section 4).

1.2. BASIC ECONOMIC VIEWPOINTS CONCERNING ENVIRONMENTAL POLLUTION

Before addressing the question of whether liability rules or safety regulation should be used to control environmental pollution, the question should be answered of whether legal rules have any function at all with respect to environmental damages and if so, what. In the economic analysis of law, legal rules are not seen as a goal in itself, but as a means to reach another goal, e.g. a reduction of environmental damages. With respect to environmental damages the function of law can be described as "internalizing the externality of environmental damages". Environmental pollution is the classic example of a negative external effect or externality[2]. If no legal rule existed, a polluter would not take into account the negative effects on the environment which he causes by his activity. Without law the externality would not be internalized.

Economists usually take as starting point for such analysis the well known Coase-theorem which states that, in a zero transactions costs, world allocative efficiency would be reached irrespective of the legal rule[3]. In Coase's world property rights are fully defined, enforced and freely transferable. The Coase-theorem could be applicable to environmental pollution cases if, for example, a polluter could cause harm only to a few well-identified victims, who were able to negotiate freely with the polluter[4]. In that case an

2. See Van den Bergh, R., "Le Droit Civile face a l'analyse economique de droit", *Ride*, 1988, 229-254. This concept of "negative externality" is also discussed in the paper by the subgroup of economists, section 2.1.
3. See Coase , R., "The problem of social cost", *Journal of Law and Economics*, 1960, 1-44; for a summary see Posner, R., *Economic analysis of law*, third edition, Boston, Little, Brown & Co., 1986, p. 42-48.
4. For applications of the Coase-theorem with respect to environmental pollution see e.g. Posner, R., *Economic analysis of law*, p. 54-57; Landes, W. and Posner, R., *The economic structure of tort law*, Cambridge, Harvard University Press, 1987, p. 42-53.

internalization would be reached through the agreement reached between the two parties to reduce the environmental damages to an efficient level. However, this Coase-theorem is only applicable if the transactions costs are equal to zero[5]. In most cases of environmental pollution the transactions costs will be prohibitive. In many cases, large number of people will have become victims of the pollution, or the whole community could suffer harm, e.g. in case of pollution of drinking water. It follows, therefore, that in the majority of environmental pollution incidents the assumptions of the Coase-theorem will not be fulfilled and an intervention of the legal system is therefore necessary to reach an internalization of the environmental da-mages.

As mentioned in the introduction, one way to reach this internalization is by using the deterrent effect of liability rules. The economic literature on tort law has demonstrated that liability rules give incentives to parties to behave carefully. Much attention has been paid to the definition of the *efficient* level of care. This is now defined as the level of care where the marginal costs of care (the costs of being more careful) and the marginal benefits (the additional reduction in expected damages) are equal[6].

This approach can also be used with respect to the damages caused by environmental pollution. Efficiency can be reached if a potential polluter can be forced by the threat of liability to make efficient investments in environmental protection equipment. The level of efficient care would thus be reflected in optimal safety techniques which would lead to an optimal reduction of the environmental damages[7].

We have now defined the goal of environmental law more closely: it should lead to an internalization of environmental damages by forcing the potential polluter to spend on efficient care. In the literature generally three instruments are listed whereby this goal can be achieved. We already mentioned two of them: safety regulation and liability rules. A third possibility is the use of taxes to force a potential polluter to invest in environmental safety techniques. Many economists have advanced a tax system as preferable from an efficiency view point[8]. This taxing system, however, has not yet been applied in many legal systems. Therefore we will now address the criteria for safety regulation versus

5. See on the transaction costs in more detail: Williamson, O., "Transaction costs economics: the government of contractual relations", *Journal of Law and Economics*, 1979, 233-261 and Williamson, O., "Contract-analysis: the transaction cost approach", in *The economic approach to law*, Burrows, P. and Veljanovski, C. (eds.), London, Butterworths, 1981, 39-60.

6. See Brown, J . P ., "Toward an economic theory of liability", *Journal of legal studies*, 1973, 323-349.

7. It should be noted, however, that these models assume risk neutrality. In this stage of the analysis we therefore neglect problems of risk and uncertainty. The level of efficient care might be higher in case of risk aversion.

8. See Klaus, J., "Umweltökonomie und Umweltpolitik", in *Ökologie, Ökonomie und Jurisprudenz*, Wenz, E., Issing, O. and Hofmann, H. (eds.), München, 1987, 10-14; Ackerman, B., Rose-Ackerman, S., Sawyer, J. and Henderson, D., *The uncertain search for environmental quality*, London, McMillan, 1974, p. 260-281; Bongaerts, J. and Kraemer, A., "Water pollution charges in three countries", *European Environment Review*, 1987, Vol. I, No. 4, p. 12-19.

liability to examine whether such regulation by government to prevent environmental damages is preferable to a simple system of liability rules. A more general overview of the economic instruments is presented in another part of this book.

1.3. CRITERIA FOR SAFETY REGULATION

1.3.1 Information asymmetry as a criterion for regulatory intervention

Information deficiencies have often been advanced as a cause of market failure and as the justification for government intervention through regulation[9]. Also, for the proper operation of a liability system, information on e.g. the existing legal rules, the accident risk, and efficient measures to prevent accidents, is a precondition for an efficient deterrence. According to Shavell, the parties in an accident setting generally have much better information on the accident risk than that possessed by the regulatory body[10]. The parties themselves have, in principle, the best information on the costs and benefits of the activity that they undertake and of the optimal way to prevent accidents. This "assumption of information" will, however, be reversed if it becomes clear that some risks are not readily appreciated by the parties in an accident setting. Therefore, for every activity the question that will have to be asked is whether either the government or the parties involved can acquire the information at the least cost.

1.3.2 Insolvency risk

If the potential damages can be so high that they will exceed the wealth of the individual injurer, liability rules will not provide optimal incentives. The reason is that the costs of care are directly related to the magnitude of the expected damages. If the expected damages are much greater than the individual wealth of the injurer, the injurer will only consider the accident as having a magnitude equal to his wealth. He will take, therefore, only the care necessary to avoid an accident equal to his wealth, which can be lower than the care required to avoid the total accident risk[11]. This is a simple application of the principle that the deterrent effect of tort liability only works if the injurer has assets to pay for the damages he causes. If an injurer is protected against such liability, a problem of underdeterrence arises[12].

9. See the basic article by Stigler, G., "The economics of information", *Journal of political economies*, 1961, p. 213 and Schwartz, A. and Wilde, L., "Intervening in markets on the basis of imperfect information: a legal and economic analysis", *University of Pennsylvania Law Review*, 1979, 630-682; Mackaay, E., *Economics of information and the law*, Boston, Kluwer, 1982.
10. Shavell, S., *Journal of legal studies*, 1984, 359.
11. Shavell, S., *Journal of legal studies*, 1984, 360.
12. Shavell, S., "The judgement proof problem", *International review of law and economics*, 1986, 43-58.

Safety regulation can overcome this problem of underdeterrence caused by insolvency.[13] In that case the efficient care will be determined *ex ante* by regulation and will be effected by enforcement instruments which induce the potential injurer to comply with the regulatory standard, irrespective of his wealth.

In that case a problem might arise if the regulation was also enforced by means of monetary sanctions. Again, if these were to exceed the injurer's wealth, the insolvency problem would remain. Hence, if a safety regulation is introduced because of a potential insolvency problem, the regulation itself should be enforced by non-monetary sanctions[14].

1.3.3 The threat of a liability suit

Some activities can cause considerable damage, but even so a law suit to recover these damages may be never brought. If this was to be the case, there would of course be no deterrent effect of liability rules. Therefore, the absence of a liability suit would again be an argument to enforce the duty of efficient care by means of safety regulations rather than through liability rules[15]. There can be a number of reasons why a law suit is not brought, even though considerable damages have been caused.

Sometimes an injurer can escape liability because the harm is thinly spread among a number of victims. As a consequence, the damage incurred by every individual victim is so small that he has no incentive to bring a suit. In particular, this problem will arise if the damage is not caused to an individual but to a common property, such as e.g. the surface waters in which each member of the population has a minor interest. In addition, a long time might have elapsed before the damage becomes apparent; in this case much of the necessary evidence may be either lost or not obtained. Another problem is that if the damage only manifests itself years after the activity, the injurer might have gone out of business.

A related problem is that it is often hard to prove that a causal link exists between an activity and a type of damage[16]. The burden of proof of a causal relationship becomes more difficult with the increasing passage of time since the damaging incident took place. Often a victim will not recognise that the harm had been caused by a tort, but might think that his particular ailment, e.g. cancer, had a "natural cause", associated with a general

13. Since we assume risk-neutrality insurance does not play a role here. If insurance would come into the picture it could overcome the problems of underdeterrence, provided that the moral hazard problem, caused by insurance, can be cured.

14. Shavell, S., "Criminal law and the optimal use of non-monetary sanctions as a deterrent", *Columbia law review*, 1985, 1232-1262.

15. Shavell, S., *Journal of legal studies*, 1984, 363.

16. See Landes, W. and Posner, R., "Tort law as a regulatory regime for catastrophic personal injuries", *Journal of legal studies*, 1984, 417.

ill health . For all these reasons a liability suit might not be brought and hence safety regulation is necessary to ensure that the potential polluter takes efficient care[17].

1.3.4 Administrative costs

When examining the pro and contra's of liability versus regulation, the administrative costs of both systems should also be compared. Liability rules are clearly costly in terms of time for both parties and in court fees. A part of these costs is borne by the whole community, such as e.g. the cost of the legal system, fees for the judges etc. Regulation produces costs for the community, including the costs of making the regulation, setting the standards, passing the statutes etc. and of subsequent enforcement[18].

In this respect the liability system seems to have an advantage: the administrative costs of the court system are only incurred if an accident has actually happened. The main advantage of the tort system is that a lot of accidents will be prevented by the deterrent effect of being held liable and having to pay damages to the victim. In case of safety regulation the costs of passing the regulation and of enforcing it are always there, whether there are accidents or not.

1.4. REGULATION IN THE PUBLIC INTEREST OR FOR INTEREST GROUPS?

Up to now we have assumed that governmental regulation is always made "in the public interest". Thus, the government always makes regulation to increase social welfare and to raise efficiency. This implies that the government would pass a regulation only if it was considered to be efficient in the terms of the criteria of Shavell mentioned above. Reality is often different. Looking at legislation one can often see regulation of activities for which such a method of control seems inefficient; in other cases there is a proper argument for regulation, but the formulation of the regulation is inefficient.

Economists of the "public choice" school have studied the phenomenon of regulation and have examined the reasons why some are inefficient[19]. In this public choice approach, micro-economic models are used to explain processes of regulation. A starting point is the assumption that the politician himself is also a *homo economicus*, who will have a preference for the introduction of regulation which will benefit those interest groups which support him. From this point of view the politician is no longer someone who is serving the public interest, but a utility maximiser serving his own interests.

17. For alternatives to liability suits see: Bocken, H., "Alternatives to liability and liability insurance for the compensation of pollution damages", *TMA* 4-87, 83-87 and *TMA* 1-88, 3-10.

18. Shavell, S., *Journal of legal studies*, 1984, 363-364.

19. See in general: Buchanan, J. and Tullock, G., *The calculus of consent*, University of Michigan Press, 1962; Buchanan, J., Tollison, R. and Tullock, G., (eds.), *Toward a theory of the rent seeking society*, University of Michigan Press, 1962.

Regulation is considered as the product of supply and demand on a political market. On the demand side we find the various interest groups who demand favourable regulation. On the supply side the politicians who wish to maximise their utility by favouring interest groups which support them politically. The product is regulation which protects interest groups in exchange for political support e.g. during the election campaign.

The reason why this wealth transfer to certain interest groups can take place can be found in two well-known economic phenomena: information- and transaction costs. The *information* costs are those incurred by the public at large in finding out the effect of the regulation. Because the wealth transferred to an interest group is derived from the general public, one individual will usually not realise that such a transfer to an interest group has taken place. Moreover, most interest group legislation is claimed to protect the general public. This is often the case with regulation which is supposed to protect consumers (and which often only protects the retailing trade)[20] or regulation which is supposed to protect victims (and which often only protects the interests of insurers of certain industries)[21].

Transaction costs are those incurred by a group in bringing individuals together to defend a common interest. Only a group which is relatively small, well organized and single issue oriented, will have low transaction costs and will be able to benefit from wealth transfers through regulation[22]. For a long time it was held that safety regulation, such as environmental legislation should not be considered as the result of lobbying by interest groups, since that kind of regulation is always issued in the general interest. However, Maloney and McCormick have shown that regulation is sometimes used as an instrument of wealth maximization, when it becomes a wealth transfer to certain interest groups[23]. They argue that if a sector of industry realizes that regulation is unavoidable, they will cooperate in the development of the regulation and try to change the contents to its advantage. A classic example is the introduction of so-called "grandfather-clauses" which stipulate that the regulation will not be applicable to firms or products which are

20. This has been proven with respect to Sunday closing regulation and "unfair trade practices" statutes. See Van den Bergh, R., "Belgian public policy towards the retailing trade", in: *Law and Economics and the Economics of Legal Regulation*, Graf von den Schulenburg, J.M. and Skogh, G. (eds.), Dordrecht, Nijhoff, 1986, 185-205. See also the article of Jaffer, Susan M., and Kay, John A., "The regulation of Shop Opening Hours in the United Kingdom", in: *Law and Economics and the Economics of Legal Regulation*, Graf von den Schulenburg, J.M. and Skogh, G. (eds.), Dordrecht, Nijhoff, 1986, 169-183.
21. This has been proven with respect to legislation concerning the liability for nuclear accidents. See Faure, M. and Van den Bergh, R., "Liability for nuclear accidents in Belgium from an interest group perspective", in: *International Review of Law and Economics*, 1990, 10, 241-254.
22. See Olson, M., *The rise and decline of nations*, 1982, 26-27; Posner, R., "Theories of economic regulation", *Bell Journal of economics*, 1974, 345; McCormick, R. and Tollison, R., *Politicians, Legislation and the economy: an inquiry into the interest group theory of governments*, 1982, p. 17.
23. See Maloney, M. and McCormick, R., "A positive theory of environmental quality regulation", *Journal of Law and Economics*, 1982, 100-108.

already in existence. Hence, the regulation can create a new barrier for market entry and so protect the existing industrial practices and products[24].

1.5. SUMMARY

Under Section 3 we have advanced several criteria which could be used to determine whether in a certain case the internalisation of damages could better be reached through the liability system than through safety regulation and vice versa. Of course, these criteria do not always provide clear arguments which favour either liability rules or safety regulation. Certain criteria such as the insolvency risk and the risk that no liability suit will be brought, clearly indicate an advantage for safety regulation. But the other criteria, such as information and administrative cost, generally favour the liability system. However, as will be indicated below (see 2.2.), in certain cases (e.g. for environmental damages) all the criteria seem to favour safety regulation. However, even if safety regulation seems to be more efficient in a specific case than liability rules, this does not mean that the regulator would opt solely for regulation and exclude the use of liability rules. One problem with regulation is that enforcement is often lacking. Therefore one can see that in most legal systems safety regulation is combined with liability rules.

After having discussed the criteria for safety regulation we indicated that this approach is not always given due emphasis, which is unfortunate given the public interest arguments in favour of regulation. Regulation often seems to favour certain interest groups. As a result of lobbying, regulation is sometimes pronounced in cases where it seems inefficient; in other cases it is the contents of the regulation that seem inefficient.

A combination of this public interest approach and the interest group theory shows that, in understanding the existence of regulation in a certain field or in examining whether regulation would be desirable in a certain field, one could first use the public interest approach to examine what kind of regulatory instrument (liability rules, taxes or regulation) would be the most efficient. If as a next stage one could discover regulations which do not fit into this public interest model, it would be helpful to look at the possible influence of interest groups in the regulatory process which might explain the existence of inefficient regulation. One should of course bear in mind that both the public interest and the interest group theory are positive economic theories, i.e.: they try to explain the existing scope of regulation, without making a judgement about whether inefficient regulation is "bad".

This part has shown that a combined Law and Economics approach can provide useful insights in the functioning of regulation. The economic theory of regulation can

24. Therefore Buchanan and Tullock argue that industries favour regulation instead of taxes, since it is easier to limit market entry through regulation (see Buchanan, J. and Tullock, G., "Polluters profits and political response: direct controls versus taxes", *American Economic Review*, 1975, 139-147); for a critique see Hahn, R., "Economic prescriptions for environmental problems: how the patient followed the doctor's orders", *Journal of Economic Perspectives*, 1989, 95-114.

provide clear criteria for regulation, which can be useful in the deregulation debate. This part also showed that a combined use should be made of several economic doctrines, such as the public interest and the private interest theory. In the next part we will examine whether a regulation of pesticides is desirable, taking into account the criteria for regulation as discussed in this part.

Part 2
The EEC Directive 80/778: Institutional analysis

2.1. INTRODUCTION

In this part of the paper we will look more closely at the EC Council Directive of 15 July 1980. We will consider the question of whether a regulation of pesticides as such is necessary, taking into account the criteria given in Section 2 below. If the answer to this question is positive, the next step is to seek the reasons given by the drafters for setting the maximum admissible concentration for pesticides in drinking water at $0.1\mu g/l$. Therefore, an overview will be given of the contents and general structure of the Directive (Section 3). An overview of the legal history will be provided (Section 4) in so far as this is useful for answering the question why the standard for pesticides has been set at this particular level. The final question to be answered is whether the chosen standard for pesticides can be considered to be efficient and, if it is not, what was the reason for choosing this concentration (Section 5).

2.2. IS REGULATION OF PESTICIDES DESIRABLE?

If one takes the criteria for safety regulation discussed in part 1 and applies them to the potential risks caused by pesticides, there is no doubt that liability rules alone are insufficient. If one looks at the first criterium, that of information costs, it must be stressed that an assessment of the risks of a certain activity often requires expert knowledge and judgement. Small organisations might lack the incentive or resources to invest in research to find out what the optimal care level would be. Also, there would be little incentive to carry out intensive research if the results were automatically available to competitors in the market: this is the well known "free rider" problem. This problem can partially be countered by legal instruments granting an intellectual property to the results of the research. However, the problem remains that it may not be possible for small companies to undertake studies on the optimal technology for preventing environmental damage. Therefore, it is often more efficient to allow the government itself to do the research on the optimal technology (e.g. in a governmental environmental research institute).[25] The results of this research can then be passed on to the parties in the market through the regulation. Hence, the setting of environmental standards in regulation can be seen as a means of passing on information on the minimal environmental technology required. Applying this more specifically to pesticides, one could imagine that it is more efficient for the government to acquire information on the amount of pesticides which represents the optimal care level, than it would be for e.g. the individual farmer to find out what additional reduction in the use of pesticides would produce an optimal reduction of the expected damages from the use of this pesticide.

Also, the insolvency argument points in the direction of regulation. Pesticides are used by farmers and other individuals or firms with assets which are generally lower than the damages they can cause by using the pesticide. In this respect it should not be forgotten that one farmer who uses a dangerous pesticide can cause harm to a very large number of individuals. The amount of damages caused by this emission can of course largely exceed his individual assets. Even if the harm is caused not by an individual farmer but by a firm, it should also be stressed that an insolvency problem remains. Indeed, most firms have become incorporated as a legal entity and therefore benefit from limited liability. Hence, the individual shareholders are not liable to the extent of their personal assets, but a creditor of the firm can only lay claim to part on all of the total assets purchased in the firm by the shareholders.

Also the chances of a liability suit being brought for damage caused by wrongful use of pesticides is naturally very low. The damage is often spread over a large number of people, who will have difficulties to organize themselves to bring a law suit. In addition, the damage could only become apparent some years after the emission of the pesticide took place. This will bring proof of causation and latency problems, which will often make it difficult for a law suit to be brought against the emitter of the pesticide.

For these reasons it is clear that some form of government regulation of the use of pesticides is necessary. This does not, however, mean that we should solely rely on government regulation and forget about the tort system where harm caused by pesticides is concerned. One reason is that the enforcement of regulation is often weak[26]. Therefore safety regulation should always be combined with liability rules. This has the additional advantage that many of the problems with the tort system can be overcome because the judge who has to decide a liability case will be able to use the regulatory standard to examine whether optimal care was taken or not. This also applies in the legal system since most tort systems hold that the violation of a regulatory norm is automatically considered

25. The assessment of optimal technology is of course easier when the pollution is caused by a specific discharge than when the pollution comes from diffuse inputs (pesticides, fertilizers), where the individual discharger may be difficult to identify.

26. Wheeler is very sceptical about the use of criminal law to enforce safety regulation (See Wheeler, M., "The use of criminal statutes to regulate product safety", *Journal of legal studies*, 1984, 593-618).

negligent or faulty behaviour.[27] Hence, the regulatory standard guides the judge in fixing the standard of care in a tort case[28].

2.3. CONTENTS AND STRUCTURE OF THE EEC COUNCIL DIRECTIVE 80/778

Let us now turn to the Directive with which it all started; the Council Directive of 15 July 1980 relating to the quality of water intended for human consumption[29].

2.3.1 The considerations

The considerations preceding the Directive state in general that "in view of the importance for public health of water for human consumption, it is necessary to lay down quality standards with which such water must comply". This statement is interesting at the outset because it makes clear that in the opinion of the drafters of the Directive the goal of the quality standards is the protection of public health, not the protection of surface waters in general.

The legal basis used for this Directive is article 100 of the EEC Treaty. This article allows the Council to approximate the laws of the various member states to Directives if there is a risk that differences in the laws could distort competition. This argument is literally referred to in the considerations preceding the Directive[30]. Apparently the Council argued that differences in the quality standards for drinking water could affect the competition between the member states. This argument seems realistic. If indeed in one member country certain pesticides were to be banned, whereas they were still allowed in other, this could have an effect on the food production costs in the different states and would thus distort a fair competition.

In addition it is stated that "the values fixed for certain parameters must be equal to or lower than maximum admissible concentration".

A final point in the considerations is the statement that the reference methods of analysis which are defined in the annexes to this Directive must be speedily adapted to scientific and technical progress.

27. Of course the violation of a regulatory norm is only one condition for liability in tort. The injurer is only bound to pay compensation if the injured can prove that he has suffered damage and a causal relationship exists between the fault and the damage.
28. See Faure, M. and Van den Bergh, R., "Negligence, strict liability and regulation of safety under Belgium law: an introductory economic analysis", *Geneva papers on Risk and Insurance*, 1987, 109-111.
29. *Official Journal of the European Community*, No. L 229/11 of 30 August 1980.
30. It states: "Whereas a disparity between provisions already applicable or in the process of being drawn up in the various member states relating to the quality of water for human consumption may create differences in the condition of competition .....".

2.3.2 The text

Let us now turn to the articles of the Directive.

Article 1 stipulates that the Directive is applicable to "water intended for human consumption". Article 2 immediately follows and defines that "water intended for human consumption" is all water used for that purpose, either in its original state or after treatment, regardless of origin, whether supplied for consumption, or whether used in a food production undertaking for the manufacture, processing, preservation or marketing of product or substances intended for human consumption and effecting the wholesomeness of the foodstuff in its finished form. In this respect the Court of Justice has stipulated[31] that "the Directive applies only to water supplied for human consumption and to water used in foodstuffs by a food production undertaking and that water from private sources of supply is excluded from its scope". Hence, one should bear in mind that the drinking water Directive does not apply to private wells.

Article 4 stipulates that the Directive is not applicable to natural mineral waters, nor to medicinal waters. Important is article 4, (2 al.), which provides that member states may not prohibit or impede the marketing of foodstuffs on grounds relating to the quality of the water used, where the quality of such water meets the requirement of the Directive, unless such marketing constitutes a hazard to public health. This is an important provision, since it means that member states cannot in principle impose requirements other than those in the Directive. It follows therefore that the set of standards in the Directive are a maximum, not a minimum. This rule can be understood if the goal of the Directive is the harmonization of laws and the desire to avoid market distortion through trade barriers. An exception is made, however, if the marketing of food stuffs would constitute a hazard to public health. In such a case a member state would be allowed to prohibit the marketing of foodstuffs on grounds relating to the quality of the water, even though that water meets the requirements of the Directive.

Another important article is article 7, which contains the duties of the member states with respect to the setting of parameters. The first paragraph provides that member-states shall fix values applicable to water intended for human consumption for the parameters shown in Annex I. We will discuss the contents of this Annex I below. However, paragraph 2 stipulates that the member states may refrain from fixing the value of parameters in respect of those for which no value is shown in Annex I, as long as these values have not been determined by the Council. Furthermore paragraph 3 stipulates that for the parameters in tables A-E of Annex I the values to be fixed by the member states must be less than or the same as the values shown in the "maximum admissible concentration" column. Hence, for these parameters the member states can deviate from

31. In Case C 42/89 of July, 5, 1990, Commission v. Kingdom of Belgium, which will be discussed extensively below.

the value in the Directive, but only if the value is less than the maximum admissible concentration in Annex I. A further duty is put on the member states in paragraph 4, which stipulates that member states shall take the steps necessary to ensure that water intended for human consumption meets at least the requirements specified in Annex I. This duty for the states is reinforced in article 8 which obliges the member states to take all the necessary measures to ensure that any substances used in the preparation of water for human consumption do not remain in concentrations higher than the maximal admissible concentration and that they do not constitute a public health hazard. Hence, article 8 explicitly refers to the substances which would be added to the water to make it suitable for human consumption.

Article 9 allows the member states to derogate from the provisions of the Directive, among others to take account of situations arising from exceptional meteorological conditions. However, paragraph 3 of article 9 provides that the derogations shall never relate to toxic or microbiological factors or constitute a public health hazard.

As far as the enforcement of the Directive is concerned, article 12 stipulates that the member states should take all necessary steps to ensure a regular monitoring of the quality of water intended for human consumption. This monitoring should take place at the point where it is made available to the user, in order to check whether it meets the requirements laid down in Annex I.

Also interesting is article 16 which provides that without prejudice to article 4 (2) member states may lay down more stringent provisions than those provided for in this Directive for water intended for human consumption. The consequence of this article is that member states are allowed to impose more stringent quality standards than those provided for in Annex I, as long as this does not mean that a state would prohibit the marketing of foodstuffs on grounds relating to the quality of the water, if the quality of such water meets the requirements of this Directive. Hence, as far as the quality of the water itself is concerned, more stringent parameters may be imposed, but not as far as foodstuffs are concerned for which drinking water is used.

As far as the time by which the Directive should be implemented is concerned, article 18 requires the member states to bring into force the laws, regulations and administrative provisions necessary to comply with this Directive within 2 years following its notification. Furthermore article 19 requires the member states to take the necessary measures to ensure that the quality of water for human consumption complies with this Directive within 5 years of its notification. This means that any new legislation which had to be passed by the member states would have to be in place within a period of 2 years, which is very short, considering that legislative processes are slow in most member states. These new laws should lead to an improved quality of drinking water which should then meet the standards of the Directive within 5 years.

2.3.3 Annexes

Several references have already been made to Annex I which contains the list of parameters. In the context of this paper the parameters in Section D concerning toxic substances are of interest. For pesticides and related products a maximum admissible concentration of 0.1 $\mu g/l$ is allowed for such substances considered separately, with the total concentration not to exceed 0.5 $\mu g/l$. The comment to this parameter number 55 stipulates that "pesticides and related products" means:

- insecticides including: persistant organochlorine compounds
  organophosphorous compounds
  carbamates
- herbicides
- fungicides
- PCB's and PCT's

So far as the reference methods of analysis for parameter 55 (pesticides and related products) are concerned, Annex III refers to the tests for other organochlorine compounds being: "gas-phase or liquid-phase chromatography after extraction by appropriate solvents and purification identification of the constituents of mixtures is necessary. Quantitative determination".

Referring back to the text of the Directive it must be stressed that pesticides are classified together with the toxic substances. Therefore no derogation of the standard in the annex of the Directive can be made by the member states.

The Directive itself is silent with respect to the question of why the parameter for pesticides has been set at 0.1 $\mu g/l$. For none of the parameters in the Directive is there any mentioning of the basis for the chosen standard. It seems nevertheless an interesting question to examine why the drafters of the Directive have set a value at a certain limit. In other words what scientific or technical knowledge did they use in setting the values given? In the following section the legal history of the Directive will be addressed to examine whether an answer to this question can be found there.

2.3.4 *Amendments*

For the sake of completeness, it should be mentioned that the drinking water Directive, discussed here, has been amended several times:

- on October 19, 1981, by Directive L 81/858, because of the entry of Greece into the European Community[32];
- on June 12, 1985, Annex I, chapter X point 1. And Annex XXXVI, chapter III point 6, because of the Accession of Spain and Portugal[33];
- on December 17, 1990, by Directive L 90/656 because of the Unification of Germany[34].

2.4. LEGAL HISTORY OF THE COUNCIL DIRECTIVE OF JULY 15, 1980

In the considerations preceding the Directive it is stated that the initiative for this Directive was related to the 1973 and 1977 programmes of action of the European Communities on the environment. Both programmes provide for a setting of standards to apply to toxic chemical substances and to bacteria that present a health hazard, which are present in water intended for human consumption. These programmes also provide for the definition of physical, chemical and biological parameters appropriate to the different uses of water and, in particular, to water for human consumption.

It is interesting to have a quick look at this first programme, the so-called Declaration of the Council of the European Communities and of the Representatives of the governments of the member states of 22 November 1973 on the Programme of Action of the European Communities on the Environment[35]. In Chapter 2 the stated aim is that standards should be established in order to limit or prevent the exposure of targets. This is put forward as a means of achieving or approaching quality objectives. With respect to the toxic chemical substances present in water intended for human consumption the programme of action mentions that a proposal from the Commission should be submitted as quickly as possible and at the latest by 31 December 1974. In addition the text explicitly provides:

32. *Official Journal*, nr. L 319/19 of 7 November 1981.
33. *Official Journal*, nr. L 302/9 of 16 June 1985.
34. *Official Journal*, nr. L 353/59 of 17 December 1990. Germany has been authorized to derogate in the sense that the obligations following from Directive 80/778 will only have to be fulfilled by December, 31, 1995.
35. *Official Journal*, nr. C 112/1-53 of 20 December 1973.

"Maximum use will be made of the results already achieved at national and international level, particularly the work done by the WHO[36].

As we know now, the deadline for submitting a proposal of 31 December 1974 was not met by the Commission because the first proposal for a Council Directive relating to the quality of water for human consumption was only submitted to the Council by the Commission on 31 July 1975[37]. This original proposal of 1975 was considerably shorter than the Directive which was finally adopted in 1980. The original proposal of 1975 consisted of only 13 articles. Remarkable though, is that the limitative standard for pesticides and related products of 0.1 μg/l can be found in this original proposal of 1975. It was included in Annex I which contained the drinking water standards, under table D "undesirable or toxic factors". However, the original proposal gives no indication of why the standard for pesticides has been set at this level. It is furthermore interesting to note that the original proposal of 1975 also contained a possibility to make derogations. A proposed article 5 in the 1975 proposal allowed member states to make provision for departures from the terms of the Directive in order to take account of *inter alia* situations arising from the nature and structure of the ground in the geographical area from which the source in question emanates. This was called an "exceptional maximum admissible concentration". Paragraph 3 of the proposed article 5 provided that in no case the departures taken by member states by virtue of that article shall exempt them from the conditions imposed for the protection of public health. However, it is important to note that the proposed article 5 did not explicitly say that these derogations can never relate to toxic or micro-biological factors, such as the present article 9, paragraph 3 does. Hence, we can conclude that, although the reasons for setting the parameter at the 0.1 μg/l level were not given, this level apparently originates from the 1975 proposal. However, the 1975 proposal still made derogations possible, even for the "undesirable or toxic factors", so long as the protection of public health was not endangered. The present text of the Directive still has the 0.1 μg/l limit but a derogation is no longer possible.

After the Commission had drafted the original proposal and had submitted it to the Council on 31 July 1975, the usual formal procedure of consultation and receiving advice took place. A first opinion was given by the European Parliament in the form of a resolution[38]. This resolution of the European Parliament, however, does not contain much useful information on the question of how the parameters in the annexes to the proposal of 1975 were set. The only thing that is worth mentioning is that the European Parliament urges the Commission to ensure that the parameters and the values attached to them are reviewed and adapted to new research results at regular intervals, at least every five years,

36. *Official Journal*, nr. C 112/14 of 20 December 1974.
37. *Official Journal*, nr. C 214/2 of 18 September 1975.
38. *Official Journal*, nr. C 28/27 of 9 February 1976.

in order to guarantee optimal health protection. In addition the European Parliament suggested a couple of amendments with respect to the text of the proposal of Directive, but these are not worth mentioning. The European Parliament made no amendment whatsoever with respect to Annex I (the annex containing the specific parameters).

Another opinion came from the Economic and Social Committee on 25 and 26 February 1976[39]. The balance of this opinion is somewhat different from that of the European Parliament. For instance, in the considerations preceding this opinion it is stated:

"Whereas in many places within the European Communities the quality of water for human consumption is satisfactory and has been so for many years".

But nevertheless they agree that "there is an increasing danger of pollution of water by toxic chemical substances and micro-biological organisms". In the general comments under 1.1 the Economic and Social Committee notes that a number of practical problems will arise if the Directive is applied in every detail in its present form. They argue that the application of all the standards referred to in the Directive will create, especially, problems for industry. The Committee considered that consultations should take place with experts from industry on the establishment of specific parameters (e.g. for zinc). In point 1.3 it is, moreover, argued that in many instances the values of the quality parameters seem to be "arbitrary and unrealistic". It is further argued that some of these parameters should be more realistic in their limiting values, so as to avoid many existing satisfactory supplies being unnecessarily classified as unsatisfactory, and to avoid subsequent heavy expenditure in seeking to make these supplies conform with "rigid and arbitrary requirements". In addition the Committee points out that "in various sectors of the foods and drinks industry, water is in fact used which does not comply with the proposed Directive, and apparently no harm to public health has resulted".

These arguments show that the Economic and Social Committee was critical of the proposed Directive, without being very specific about the parameters to which the Committee took exception. In point 1.7 the Committee notes with satisfaction that the standards laid down in the annexes relate broadly to those of the World Health Organization (WHO). But there is also criticism: "However, in many instances the standards are much stricter and the Committee wonders what justification there is for this".

In addition to these general comments the Committee also formulates a number of specific comments. In point 2.3.1 the Committee makes a general criticism about the parameters referred to in Annex I, in that according to the Committee these parameters are much too detailed. The Committee suggests that it would be more practical to confine the Directive to specifying maximum limits for those substances known to be toxic and for

39. *Official Journal*, nr. C 131/13 of 12 June 1976.

which control is therefore essential and to specific guidelines for all those parameters where there is no evidence of a health hazard.

To sum up, the Economic and Social Committee formulated a number of comments on the proposal of the Directive. Most of these comments were of a rather general nature and indicated a fear that the values in the parameters were arbitrary and unrealistic. However, detailed comments on all the parameters in Annex I were not made. It is, moreover, interesting that the Economic and Social Committee, just as the programme of action of 1973, indicated that the standards should be set taking into account the values laid down by the World Health Organization. Apparently the Committee noted that in some instances the standards were stricter than the WHO-standards, without any justification. However, the Directive of 15 July 1980 did not take the criticisms of the Economic and Social Committee into consideration since no specific reference can be found in the considerations preceding the Directive of why in certain cases the WHO-standard has not been followed.

The most important conclusion from this overview of the legal history of the Directive is probably that it provides no information whatsoever as to why the drafters of the Directives did not follow the WHO-standards for all parameters. This is especially noteworthy since reference to the WHO-parameters was made both in the programme of action of 1973 and in the resolution of the Economic and Social Committee.

2.5. EFFICIENCY OF THE 0.1 MG/L MAXIMUM ADMISSIBLE CONCENTRATION: PUBLIC INTEREST VERSUS INTEREST GROUP ANALYSIS

In Section 2 we concluded that a case can be made for some kind of government regulation of pesticides, taking into account the criteria given in chapter 1 for environmental regulation. A more interesting question is whether the particular norm chosen for pesticides of 0.1 μg/l is efficient.

This is a question which has been addressed by Premazzi/Chiaudani/Ziglio[40]. This study seems to indicate that the 0.1 μg/l in the Directive and especially the ban of atrazine in Italy is inefficient. It suggests that the WHO-guideline values for pesticides are more realistic. Given this previous research it seems likely that the 0.1 μg/l standard of the EC Directive of 15 July 1980 is inefficient. If this research has proven that the parameter which has been chosen in the Directive is inefficient, the question of course arises of what the reasons for this inefficiency can be.

40. See Premazzi/Chiaudani/Ziglio, 164. The paper by the subgroup of economists mentions that the WHO guideline for atrazine is 2 μg/l. (see Section 4.1).

In this respect a public choice analysis of the setting of parameters in the Directive seems appropriate[41]. It is indeed remarkable that the EC did not follow the WHO-parameters, among others for pesticides. It would thus be important to investigate more closely the dynamics of the setting of parameters at the EC level. At first glance it seems likely that certain interest groups have played an important role in the legislative process in order to obtain this tight standard for pesticides which deviates from the WHO-standard. The difficulty is, however, that the formal texts of the Directive provide no information at all as far as the setting of standards is concerned. Therefore an investigation into the dynamics of the setting of standards from a public choice perspective can only be conducted if field research can be done e.g. through interviews with members of the EC Parliament or with organized interest groups. On the other hand the question also arises whether such a detailed public choice analysis would be useful at all, since experts who followed the standard setting process with respect to the Directive closely informed me that there might be a relatively simple explanation for why the WHO-standard was not followed[42].

At the time when the EC "Environmental Directives" were discussed in the seventies, the persistent organochlorine pesticides (such as DDT) were receiving their maximum adverse publicity. It is therefore understandable that there was pressure to ensure that public drinking water should contain no pesticides. For practical reasons the standard was set at the minimum concentration of organochlorine pesticides that could be detected by the analytical methods available at that time[43]. The reason why the pesticides were not all listed separately in the annex to the Directive was that the list of parameters was already lengthy now. Given the public pressure for "clean" drinking water it would not have been possible to increase the proposed standard in respect of certain pesticides for the reason that the levels present were harmless on the basis of WHO-criteria.

Also, there was no evidence that the proposed standard was in danger of being exceeded because for many pesticides the level of analytical detection was not sufficiently sensitive. Therefore, the argument that groundwater exceeded the proposed EC standard, but still met the WHO-standards and so were safe for human consumption, was not available at that time. This was the only pressure that could have influenced the negotiations of the Directive to amend or qualify the standard, in the face of the strong political pressures (post 1972 Stockholm conference) for its adoption. Hence, the drafters of the Directive apparently never aimed at following the WHO-standard. The present problem has arisen because the increased sensitivity of modern analytical techniques for

41. See on the influence of pressure groups on EC regulation e.g. Mazey, S. and Richardson, J., "British Pressure Groups in the European Community: The Challenge of Brussels", *Parliamentary Affairs*, 1992, 92-107.

42. The following is based on information provided by Dr. Jan C. Bongaerts of the European Institute for Environmental Policy and Richard Lloyd, who had both personal insight into the legislative history of the Directive.

43. See Premazzi, Chiaudani, Ziglio, 1989.

a wide range of pesticides has shown that in many groundwaters this very stringent standard has been breached. It seems, however, politically difficult for the EC to relax or modify the standard.

2.6. RELATIONSHIP WITH OTHER EC DIRECTIVES

Although this paper mainly focuses on the drinking water Directive (80/778), we should mention that there exist of course other Directives which directly or indirectly also aim at the reduction of the pollution of waters intended for drinking water. The drinking water Directive is certainly the more important one for this research since it contains binding parameters for pesticides. It is, however, for sake of completeness, useful to take a brief look at these other Directives as well.

One Directive to be mentioned is Directive 80/68 of December, 17, 1979 concerning the protection of the groundwater against pollution through the emission of certain dangerous substances[44]. The goal of this Directive is the prevention of the pollution of the groundwater through substances listed in Annex I and II. Emissions of substances listed in Annex I are always prohibited (Article 4).

Article 3 therefore forces the member states to take all measures to prevent the emission of Annex I substances in the groundwater and to reduce the emission of Annex II substances in order to reduce the pollution of the groundwater by these substances.

The difference between this groundwater Directive and the drinking water Directive discussed earlier is that the groundwater Directive aims at a reduction of substances listed in annexes I and II in the groundwater, irrespective of the use of this groundwater as drinking water or not. Moreover, the groundwater Directive is input-oriented and only forces the member states to take measures to prevent inputs from pollutant substances into the groundwater. There are not, as is the case in the drinking water Directive, emission standards which set specific parameters with a "maximum admissible concentration"[45]. Several implementation problems have arisen with respect to this groundwater Directive. By decision of 17 September 1987[46] the European Court of Justice held that a substantial part of this groundwater Directive had not been implemented correctly in the Netherlands[47]. The problem is that the duties of the member states under this groundwater Directive are formulated so broadly that some authors hold that it is very difficult to give

44. *Official Journal* L 20/43 of 26. January 1980.
45. See on this groundwater Directive also Kohll, N.E.M., "De uitvoering van EG-milieurichtlijnen, toegespitst op milieugevaarlijke stoffen" in: *De uitvoering van EG-milieurichtlijnen toegespitst op milieugevaarlijke stoffen*, Zwolle, Tjeenk Willink, 1989, 71-95.
46. Case 291/84 of 17 September 1987.
47. See Brinkhorst, L.J., "De uitvoering van EG-milieubeleid in Nederland: het standpunt van de Europese Commissie" in: *De uitvoering van EG-milieurichtlijnen, toegespitst op milieugevaarlijke stoffen*, Zwolle, Tjeenk Willink, 1989, 8.

a full implementation to this groundwater Directive[48]. By decision of 28 February 1991 the Court of Justice also held that Italy had violated its community-obligations by not implementing groundwater Directive 68/80 in time[49].

This groundwater Directive of 1979 does not set specific parameters with quality objectives for a certain component of the environment (such as the drinking water), but focuses more on a protection of the environment through a limitation or prohibition of the input of certain substances. In this respect the important Directive 76/464 of 4 May 1976 should certainly also be mentioned[50]. This Directive aims at the prevention of the pollution of the aquatic environment through certain dangerous substances. This Directives provides a general framework, which has been worked out in other Directives. For groundwater article 4 of the Directive 76/464 provided for a duty for the member states to prohibit all emissions of substances of Annex I and to restrict the emission of substances listed in Annex II. However, article 4 also provided that this Directive 76/464 would no longer apply to groundwater as soon as a specific Directive would enter into force for groundwater, which was the case with the groundwater Directive 80/68 of 17 December 1979.

In addition to the mentioned Directives, which aim at the protection of the water quality through a restriction of certain emissions, Directive 75/440 of 16 June 1975 is again concerned with the quality required of surface water intended for the abstractions of drinking water[51]. This Directive concerns the quality requirements which surface fresh water intended for drinking water must meet. The Directive sets certain parameters in the annexes and forces the states to take all necessary measures to ensure that surface water conforms to the values laid down. The Directive divides the surface water into three categories, A1, A2 and A3, which correspond to a standard method of treatment. The Annex II also sets parameters for pesticides. For category A1[52] the standard is 0,001 mg/l; for category A2[53] the parameter is 0,0025 mg/l and for category A3[54] the parameter is 0,005 mg/l. Hence, in this Directive we find, as in the drinking water Directive, specific parameters. However, in this case the standards apply to the surface

48. See Donkers, R.H., " De uitvoering van EG-milieubeleid: het standpunt van de overheid", in: *De uitvoering van EG-milieurichtlijnen, toegespitst op milieugevaarlijke stoffen*, Zwolle, Tjeenk Willink, 1989, 19.

49. Case C 360/87. See Eeckman, P. and Vanheer, H., 1992, 1354.

50. *Official Journal*, L 129/23 of 18 May 1976.

51. *Official Journal*, L 194/26 of 25 July 1975.

52. This is simple physical treatment and disinfection, e.g. rapid filtrations and disinfection.

53. This is normal physical treatment, chemical treatment and disinfection, e.g. pre-chlorination, coagulation, flocculation, decautation, filtration, disinfection (final chlorination).

54. This is intensive physical and chemical treatment, extended treatment and disinfection e.g. chlorination to break-point, coagulation, flocculation, decautation, filtration, adsorption (activated carbon), disinfectation (ozone, final chlorination).

water which is, after treatment, used for drinking water, whereas the parameters in the drinking water Directive apply directly to the drinking water quality.

This conglomerate of various Directives does not make it easy to discover a clear line in the EC policy with regard to water quality. Several authors therefore argue that we need a more integrated approach of EC legislation[55].

Part 3
The implementation and Enforcement of the EC Directive

3.1. INTRODUCTION

In this part of the paper we will address the implementation of the EC Directive and examine some aspects of its enforcement. Section 2 will address some relevant consequences of the Directive. This section will especially focus on the duty of the member states to implement and enforce the Directive and will address the issue of a possible liability to the state if implementation is incomplete or even totally lacking. Section 3 will focus more closely on the implementation of the drinking water Directive in Italy. In that respect the various Italian derogations from the Directive will be discussed, but attention should also be paid to the question whether the quality of drinking water should be controlled by an input control (controlling emissions) or by an output control (controlling the water quality). The fourth section focuses on the sanctions and enforcement. First, a few general comments on the role of criminal sanctions in general will be made; next, the sanctions in the Italian legislation will be discussed. Section 5 will finally address the Italian ban of atrazine under EC Law. In this respect specific questions arise such as whether a prohibition of the use of atrazine can cause a distortion of competition between the member states and whether a duty to register pesticides before they can be used constitutes a possible violation of art. 30-36 of the EEC Treaty.

3.2. LEGAL CONSEQUENCES OF THE DIRECTIVE

In this section we will examine what the consequences of a Directive are. The main consequence is a duty to implement the Directive. Therefore we will focus on the possible liability of a State for failing to implement a Directive and on the question whether a Directive has binding force towards citizens. In the next section we will address several possible implementation policies with respect to the drinking water Directive and we will look more closely at the implementation in Italy.

55. See Donckers 1989.

The member states have a duty to take all general and specific measures to ensure conformity with Directives[56]. This duty follows from art. 5 EEC Treaty (Community solidarity) juncto art. 189 EEC treaty which obliges the member states to achieve the results of the Directives.[57] This duty to implement Directives does not only bind the national legislator, but also the courts which have to apply legislation in the light of the Directives[58]. This duty applies to courts in an anticipatory way, i.e. even before the time limit for the implementation of the Directive has expired. This means that the national legislation should be interpreted in conformity with EC Law.

The situation is somewhat different, however, when there is a clear conflict between EC Law and national legislation, caused by a failure to implement. In such a case criminal liability cannot be exclusively based on an unimplemented Directive[59]. An interesting example of this principle can be found in a case which concerned Directive 80/777 on the exploitation and marketing of natural mineral waters. Although this Directive should have entered into force in July 1984, the Directive was not implemented in Netherlands law.[60] However, a Dutch public prosecutor sued a café which had allegedly stocked for sale and delivery mineral water which was unsuitable for consumption. The public Prosecutor based his law suit on the unimplemented Directive. The Court held that the duty of national courts to interpret national legislation in conformity with EC Law, is limited by general principles, such as the non-retroactivity of the criminal law. Hence, criminal liability cannot rest solely on an unimplemented Directive[61].

A first conclusion therefore is that a State has a duty under EC Law to implement the Directive, and that a State can incur EC liability for failing to implement a Directive. Citizens, on the other hand, cannot incur criminal liability based on an unimplemented Directive. A citizen can, however, invoke the provisions of a Directive to its advantage, if the Directive has not been implemented in time, provided that the Directive in question has direct effect[62].

Before focusing more closely on this State-liability for non-implementation, it is worthwhile to mention that the duty to implement also brings about a duty for the State to impose sanctions on the violation of legislation implementing the Directive. The EC Court of Justice stated in several cases that although the member states are free to chose the

56. Sevenster, H., "Criminal Law and EC Law", in: *Common Market Law Review*, 1992, (29), 44.

57. See Lauwaars, R. and Timmermans, C. in: *Europees Gemeenschapsrecht in kort bestek* , Wolters-Noordhoff Groningen 1989, 33 and Geelhoed, L., Kuiper, M., Braakman, A., Vlas, P., in: *Doorwerking van het Europees Recht in de nationale rechtsorde*, Koninklijke Vermande bv, 1986, 5.

58. Sevenster, 1992, 44, see case 14/83, Van Colson and Kamann, 1984 ECR 1891.

59. Sevenster, 1992, 45.

60. The Directive was not implemented in Netherlands Law until the adoption on 26 June 1985 of the Netherlands Decree on Mineral Water (Staatsblad, 422).

61. Case 80/86 Kolpinghuis Nijmegen, 1987, *ECR* 3969.

62. Sevenster, 1992, 41, see also Case 8/81 Becker, 1982, *ECR* 53. This "direct effect" means that the provisions of the Directive should have self-executive power.

regulatory instruments to implement a Directive, the sanctions should be effective, proportional and discouraging.[63] Hence, States are not forced to introduce criminal sanctions and may choose civil liability as sanctions, provided that the sanction is proportionate to the damage suffered[64]. However, in one case, the Court held that Greece had not fulfilled its EC obligations "by failing to institute criminal or disciplinary proceedings"[65]. It is anyway clear that the Court of Justice does not accept a mere symbolic implementation which is not backed by effective and deterrent sanctions.

The same is true as far as the supervision of the EC Court of Justice on water quality Directives is concerned. The obvious problem with these kinds of Directives is that it seems relatively easy for a State to implement a Directive which contains certain parameters concerning the quality of e.g. drinking water, but that it is much harder to reach the goals set in practice as well. It has become clear, however, in a recent case that also in this area the EC Court of Justice requires more than an implementation on paper: the State can also incur EC liability if the goals set, e.g. concerning the drinking water quality, cannot be reached in practice.

A case concerning the Directive 80/778 which we are discussing in this paper, provides a nice illustration of this tendency. Belgium was prosecuted by the Commission for not having implemented Directive 80/778 correctly. The Belgian Royal Decree of 27 April, 1984[66] subjected derogations to conditions that were less strict than those laid down in the Directive. Moreover, it also appeared that the drinking water in Verviers, considerably exceeded the parameters allowed in the Directive for lead. The Belgian government had replied that it needed more time for the construction of the water treatment plant to remove the lead. This defence was rejected. Hence, Belgium was found to have acted contrary to its EC obligations, both because it had not implemented the EC Directive correctly and because the Verviers water supply did not accord with the parameters of the Directive on account of its excessive lead content. Moreover, no application for a derogation was made either in good time or in due form[67].

Jans, who wrote a comment to this case, points out the fact that in this case Belgium is held liable not only for a formal breach of its obligations under EC Law, but also because the drinking water in the city of Verviers did not meet the requirements of the Directive. According to Jans it was the first time that the Court of Justice reached such a decision. He predicts that in the future it can be expected that the Commission will prosecute more often States that have formally complied with EC Law by implementing

63. eg Case 68/88, Commission vs Greece, 1989, *ECR* 2965. See also a recent note of the Dutch Minister of Justice of 19 March 1992 concerning the relationship between EC law and Dutch criminal law: Tweede kamer, 1991-92, 22300 VI, nr. 39.

64. Sevenster, 1992, 50.

65. Case 68/88, Commission v. Greece, 1989, *ECR* 2965.

66. *Moniteur belge* of 6.7.1984, 9860.

67. EC Court of Justice, case C 42/89, of July, 5, 1990, *Milieu en Recht*, 1991, 155-158.

a Directive, but where the actual state of the environment does not meet the criteria of the Directive[68]. This will be highly important, especially since the Treaty of Maastricht of 7 February 1992 allows the Commission to ask the Court of Justice to condemn a member state which has not fulfilled its community duties to pay a penalty payment[69].

The consequence of this recent case-law for this atrazine case is obvious: to assess whether Italy has fulfilled its community obligations one should not only look at the formal regulation of implementing the Directive, but also at the values occurring in practice for all the parameters of the annexes to the Directive. If these parameters are violated, Italy can still be held liable under EC Law, notwithstanding the formal implementation.

3.3. IMPLEMENTATION OF THE DIRECTIVE

A basic question is of course how a Directive that imposes certain quality objectives should be implemented. This question becomes a particular important one now that the EC Court of Justice apparently enforces these drinking water standards upon States. Hence, a mere formal implementation is not sufficient and the question how State regulation can be used to achieve the given goals becomes very relevant.

3.3.1 Input versus output control

Two regulatory approaches. The major distinction one can make in this respect is a difference between input and output monitoring[70]. In case of drinking water quality regulation this distinction can relatively be made easily. Assuming that there are no other regulatory constraints than a legal duty to reach a certain quality for the water which reaches the consumer, a regulator could either regulate the quality of this end product or regulate the various inputs that influence the water quality. Of course, in practice regulation is often a mix between input and output control, but it seems useful to pay attention to the differences between both systems and to their comparative benefits.

The EC drinking water Directive 80/778 we are discussing in this paper is mainly output oriented: it is interested in the result, the quality of drinking water and leaves it to the member states to choose the appropriate means and methods to reach this goal. A first approach could therefore be to concentrate merely on the output and to invest substantially in water treatment equipment. In this respect we can refer to the literature which examined

68. Jans, Comment on Court of Justice, case C 42/89 of July, 5, 1990, *Milieu en Recht*, 1991, 158.
69. Art. 171 of the Treaty on the European union.
70. See on this choice the article written by Wittman, D., "Prior Regulation versus Post Liability: the choice between input and output monitoring", *Journal of Legal Studies*, 1977, 193-211.

the various treatment procedures and discusses their relative efficacy[71]. The advantage is that output control does not merely focus on one compound (as input control does). Water treatment in principle removes all pesticides without discrimination. A problem with this approach is, however, that it might be very costly to remove all pesticides, as the Directives requires. Moreover, it would also be unclear what incentives the emitters of pesticides would have to reduce the use of pesticides if all the costs would fall on the water treatment company. A solution to the latter problem could be found if the sources of the pollution could be easily identified and if the waterworks are in a position to monitor the polluter.

An obvious alternative would be an input control. Depending on what the most serious pollutant, is this would imply e.g. a control of the pesticides used in agriculture.

In principle one could make a simple cost-benefit analysis of output control on the one hand and input control on the other and choose the most effective one. The problem is, however, that many other factors play an important role as well, such as e.g. the availability of less harmful substitutive pesticides with similar agronomic qualities as the prohibited ones and the question whether or not the presence of a pesticide can indeed be attributed to one single source.

An advantage of output regulation is that it also catches the pollution caused by diffuse inputs. If an input regulation focused merely on e.g. pesticides from agricultural sources, and if pesticides also reached the groundwater from non-point sources, a regulation which prohibits only the use of a certain pesticide in the agricultural area would not necessarily achieve the goals set in the Directive. This could be a problem in the Italian case where the atrazine found in groundwater comes also from non-agricultural sources[72]. If the pollution of the groundwater had only one source which is easily traceable and if only this groundwater were used for drinking water, input monitoring could be preferred. The case would be different if drinking water were extracted from the surface water and these surface waters were polluted by many diffuse sources.

A complicating factor in case of groundwater control is, however, that the desirability of input or output control for pesticides will also depend on the persistence in the soil on which they are applied and the time it takes before they reach the groundwater, if at all. If the regulator is interested only in the output, then the location of the use of pesticides might become very important. If, e.g. on a clay-subsoil, a pesticide could never reach the groundwater, it would make little sense to prohibit the use of pesticides there by input control. The time aspect is of course important because of the problem of historic pollution. If it takes a very long time for a pesticide which was used in agriculture to

71. See the paper presented by C. Nurizzo at the May 1991 Meeting in Milan and Bongaerts, J. and Kraemer, R., *The Law and Economics of water works: Meeting the Increasing Cost of Water Production with Deteriorating Quality of Ground Water*, Berlin, September 1989.

72. One should note, however, that the use of atrazine by farmers is by far the dominating source of atrazine in groundwater in the Po valley.

actually reach the groundwater, a prohibition of its use (assuming that this would in certain circumstances be the most effective solution) might only become effective after many years. This means that such an input control should in that case be combined with an output control, at least for a certain period of time[73].

Monitoring and enforcement. There are of course differences between input and output control as far as monitoring and enforcement is concerned. Enforcement of quality standards requires, in general, considerable environmental monitoring.

If only output monitoring is used, the control is in fact directed to the waterworks that deliver the drinking water to the consumer. An input control is directed more towards the sources of the pollution, the emitting industry, or the farmers. Hence, this causes important differences for the use of sanctions and especially criminal sanctions. Inputs can be controlled e.g. by a prohibition of discharging certain harmful pesticides. Such a prohibition could be enforced by (the) criminal law. Usually the legal system either directly prohibits the use or discharge of a certain substance or it regulates the emission of the substance through the conditions in a licence (an emission permit). Usually the law punishes the person(s) who violate(s) the statutory prohibitions or the conditions of the permit. Depending upon the legal system, in case of a violation, a sanction can be imposed upon the individual or upon the firm. In this case the criminal sanction is directed towards the instigator of the pollution (industrial or agricultural), which should deter him from emitting those forbidden substances.

In the case of output control, however, the function of sanctions is a different one. One possibility is for the administrative authority, that sets the emission standards in the licence, to take into account the quality objectives. In that case a violation of these emission standards leads again to criminal liability. Another possible liability that we have already discussed above is the State Liability for violation of the standards laid down in a Directive. A final solution, which is being discussed much more, especially now in German legal doctrine, is the liability of the waterworks. Indeed, it is the waterworks that are responsible for complying with the parameters of the Directive. Hence, a violation of these standards through a lack of adequate water treatment could lead to a criminal liability

73. Also Vogelezang-Stoute and Matser argue that an input control through registration of pesticides has so many flaws that a registration scheme alone is not sufficient (Vogelezang-Stoute, E.M. and Matser, E.J., 1990, 243-262).

of the waterworks. Depending upon the legal system[74] this could either be a liability of the company, or of the responsible employee, or of both[75].

Distributional effects. One can also point to distributional differences between input and output control. If pesticides are only regulated through output control, this basically means that the waterworks will be liable for treating the groundwater before it is used as drinking water. The costs of treatment and thus of the use of pesticides will thus be borne by the waterworks. The costs will be passed on to the consumer of drinking water by increased prices[76], unless the government subsidizes the waterworks. In the latter case the costs of the use of pesticides are spread over all taxpayers. In the case of input control, the industry or the agricultural user of pesticides will bear the costs of having e.g. to use other means to protect the environment. Unless regulation provides otherwise, these extra costs of input control will of course be passed on to consumers as well. In many cases these might be the same consumers as the ones who buy the drinking water, so that the redistributive effects are in the end minimal.[77]

3.3.2 Registration of pesticides

The previous section made clear that no obvious choice between input or output control can be made. The Italian case under discussion is a particular interesting one from an institutional viewpoint, since, as we will discuss in the next section (c), Italy apparently preferred a combination of input and output control. It implemented the parameters for drinking water as laid down in the EC Directive, but at the same time it prohibited the use of particular pesticides such as atrazine. It is a separate question whether the prohibition

74. Most legal systems do not accept that legal entities can commit a crime, but there are some exceptions, for instance in art. 51 of the Dutch Criminal Code. In Germany criminal liability is only accepted for the "Ordnungswidrigkeiten" (administrative offenses). See paragraph 30 of the Ordnungswidrigkeitengesetz.

75. In Germany many authors plead in favour of a criminal liability for governmental officials who caused environmental crimes. See e.g. Keller, R., "Zur strafrechtlichen Verantwortlichkeit des Amtsträgers für fehlerhafte Genehmigungen im Umweltrecht", in: *Festschrift für Kurt Rebmann zum 65. Geburtstag*, München, Beck, 1989; Schünemann, B., "Die Strafbarkeit von Amtsträgern im Gewässerstrafrecht", *Wistra*, 1986, 235-246; Winkelbauer, W., "Die strafrechtliche Verantwortung von Amtsträgern im Umweltstrafrecht", *NStZ*, 1986, 149-153. For an Economic Analysis of this tendency see Faure, M., Oudijk, J. and Koopmans, I., "Ökonomische Analyse der Amtsträgerstrafbarkeit. Eine Skizze strafrechtlicher Steuerung von Umweltdelinquenz", *Zeitschrift für Wirtschaft-Steuer-Strafrecht*, 1992, 121-129

76. Of course this possibility of passing on the extra costs along the consumers could depend on the market structures.

77. At first sight it seems shocking to let the waterworks pay for treatment instead of the polluting industry. It looks like a violation of the polluter pays principle. But if these costs can be passed on and there is no difference in consumergroups, the distributional differences might not be that important.

of selected pesticides meets the EC requirements if as to at the same time the substitute pesticides are not controlled. It is interesting, however, that Italy combined input and output control. This is also the EC policy now. The EC regulated the quality of drinking water in Directive 80/778, as discussed in detail in Chapter III. But another Directive of July 15, 1991[78], 91/414 now regulates the marketing of "plant protection products"[79]. Article 4 of this new Directive stipulates that a pesticide can only be allowed on the market, if it has been established that the pesticide has no harmful effects on the groundwater.[80] Most EC member states already had such an input regulation, in the simple sense that a pesticide could only be put on the market if it was registered. For instance, in the Netherlands the pesticides Act of 1962[81] provides that a pesticide can only be marketed if research has demonstrated that the pesticide will not cause harm to human health or to the environment, if this harm would be unacceptable[82].

Much case-law has been developed with respect to the question whether a certain risk or harm e.g. to the groundwater is acceptable or not.[83] Interestingly enough one of these cases exactly deals with atrazine. The Dutch Minister of agriculture had decided to allow the use of atrazine until March 1, 1992, but to prohibit the marketing of atrazine after this date. To prohibit a further use of atrazine the administration had argued that pesticides are not allowed in drinking water at higher quantities than 0,1 μg/l. For that reason, so they argued, groundwater should have the same qualities. Since atrazine can still be found at quantities higher than 0,1 μg/l in groundwater,[84] its future marketing should be prohibited. Thus one can note that the administration uses the 0,1 μg/l norm from Directive 80/778 (an output norm) to justify the refusal of a continued registration of atrazine (an input control). The case was brought to the administrative court, which held that the Netherlands still had time until July 15, 1993 to implement the new Directive concerning the placing of plant protection products on the market. Until that time the

78. *Official Journal*, L 230 of August, 19, 1991.

79. See for a comment on a draft of this Directive: Vogelezang-Stoute, E.M. and Matser, E.J., *De toelating van bestrijdingsmiddelen. Milieu tussen wet en beleid*, Universiteit van Amsterdam, Centrum voor Milieurecht, 1990, 50-51.

80. This Directive has to be implemented before July, 15, 1993.

81. Act of July 12, 1962, amended by the Act of June 5, 1975. See also: Sevenster, H., *De gevolgen van de voltooiing van de interne markt voor het nationaal milieubeleid*, nr. 1990/2, 116-119.

82. See also: Vogelezang-Stoute, E.M., Simons-Vinckx, M.W.L., De Bestrijdingsmiddelenwet, Procedures en beoordelingscriteria bij de toelating van bestrijdingsmiddelen, Vereniging voor Milieurecht, 1992-2, 17-34, Verslag van de 35e ledenvergadering van de Vereniging voor Milieurecht, *Milieu en Recht*, 1992, 403-405

83. For an overview see: Vogelezang-Stoute, E.M., "Van Vaponastrip tot Atrazin. Jurisprudentie van het College van Beroep voor het bedrijfsleven inzake de toelating van bestrijdingsmiddelen", *Milieu en Recht*, 1992, 231-242.

84. Recent research indicates that in 1990 in the Netherlands in 1990 the "waterschappen" used 37 kilo's of atrazine (De Vries, P.J.R. and Wilpshaar, J., "Gebruik van bestrijdingsmiddelen door waterschappen", *Waterschapsbelangen*, 1992/2, 43-49

registration can according to the Dutch Act of 1962 only be refused if the product would have *unacceptable* harmful effects. This had not been examined by the administration and therefore the administrative decision to prohibit the marketing of atrazine after March 1, 1992, was declared void.[85]

This case illustrates that the administrative authorities can use input control by registration in order to reach the drinking water parameter, set in Directive 80/778. It also shows that Italy was maybe one of the first countries to simply ban atrazine, but that a discussion on its harmful effects has taken place in other countries such as the Netherlands as well. Moreover, through the new Directive of July 15, 1991 concerning the placing of plant protection products on the market, input controls through registration schemes will become increasingly important. The relationship between this new Directive and the drinking water Directive 80/778 would therefore merit further research. In particular, there may be a need to resolve the apparent anomalies between the standards set for pesticides residues in food in order to protect human health, and those set in the drinking water Directive[86].

3.3.3 Implementation by Italy

The regulatory framework in Italy is complex because of a combined use of input and output control. We will first give a brief overview of the evolution of the Italian legislation implementing the Directive.

Chronological overview of the regulations[87]

- 15 July 1980:
  EEC Directive 80/778[88]
  The formal date for implementation by the member states was July 17,1982. The member states had to ensure that by August 1985 drinking water met the values for parameters listed in Annex 1.

85. College van beroep voor het bedrijfsleven, 24 januari 1992, *Milieu en Recht*, 1992, 247-251.
86. Richard Lloyd informed that for example the permitted level of tecnazene in potatoes (used to control Jungal attack during storage) is 5 mg/kg. A simple calculation shows that a person would have to drink 1½ liters of water containing 0,1 μg/l tecnazene a day for 45 years to take the same amount of tecnazene as that (at its maximum permitted level) in 500 g. of potato.
87. Not being an Italian lawyer, I relied upon the information on regulatory aspects given in the Milan *Report on Ecotoxicological aspects of herbicide contamination of groundwater in Italy*. Moreover, Marco Vighi was so kind to send me the texts of the relevant Italian regulations and Sylvie van Lammeren translated them for me. We do not claim, however, that this is a complete overview of all the applicable regulations. Also the paper by the subgroup of the economists contains a chronological overview (see Section 4.1).
88. *Official Journal* L 229/11 of 30 August 1980.

- 8 February 1985:[89]
  This DPCM N.41 implements Directive 80/778 into Italian law. It merely states the implementation of the Directive into Italian Law, but contains no executive measures, nor any enforcement mechanism or regulation.
- 25 June 1986:[90]
  Order of the Italian Minister of Health with effect to 31 December 1986 which authorized the use for the supply of water intended for human consumption of water which contained residual atrazine not in excess of 1.0 μg/l in geographical areas defined by special order of the President of the competent regional Council.
- 7 October 1986:
  Order of the President of the Piemonte Regional Council authorizing until 31 December 1986 the use of water for human consumption in which the atrazine concentration did not exceed 1,0 μg/l, for the Piemonte Region.
- 22 December 1986:
  Order of the Italian minister for Health renewing the order of 25 June 1986 until 31 December 1987.
- 30 December 1986:
  Order of the President of the Piemonte Regional Council renewing the order of 7 October 1986 until 31 December 1987.
- 3 April 1987:[91]
  Order of the Italian Minister for Health authorizing until 31 march 1988 the use, in specific geographical areas of water in which residual atrazine and molinate did not exceed the limits of 1,7 μg/l and 6,0 μg/l respectively.
- 17 April 1987:
  Order of the President of the Piemonte Regional Council extending the orders of 7 October 1986 and 30 December 1986 until 31 March 1988[92].
- 24 May 1988:[93]
  This DPR nr. 236 executes Directive 80/778 into more detail. It contains detailed administrative provisions and criminal sanctions in case of non-compliance. These will be discussed below.
- 22 September 1988:
  Judgement of the Court of Justice answering the question by the Pretore di Torino.

89. Decreto del Presidente del Consiglio dei Ministri 8 Febbraio 1985, Gazetta Ufficiale n. 108 del 9 Maggio 1985.
90. Information from the Report before the Hearing concerning case 228/87, Pretura unificata di Torino v. X. discussed below.
91. Ordinanza 3 aprile 1987 n.135, Gazetta Ufficiale 6 aprile 1987.
92. Although the Ministerial order of 3 April 1987 gave the Piemonte region the possibility to increase the limit for atrazine to 1,7 μg/l, the order of 17 April 1987 did not use this option and extended the previous orders in which the limit was set at 1,0 μg/l.
93. Decreto del Presidente della Republica 24 maggio 1988, n. 236, Gazetta Ufficiale n. 152 del 30.6.1988.

- 14 February 1989:
  Decree of the Italian Minister of Health to set the maximum level for atrazine at 0,8μg/l for a maximum period of 24 months.[94]
- 21 March 1990:
  Prohibition to sell and use alachlor and atrazine in Italy until 31 December 1990.
- 5 April 1990:
  Italian Act on urgent measures for quality improvement and for preventing water pollution.
- 2 August 1990:
  Regulation to carry out DPR 24 May 1988
- 6 February 1991:
  Ordinance renewing the prohibition of 21 March 1990 until 31 December 1991
- 15 July 1991:[95]
  Council Directive concerning the placing of plant protection products on the market.

Derogations. This overview of the Italian regulations indicates that Italy allowed for derogations from the values set in the annexes to the Directive, at least for some geographical areas. Also, for countries other than Italy, it became clear that it was scarcely possible to reach the drinking water values, set in the Directive, by August 1985. This poses the question of the conditions under which derogations are possible, especially for the 0,1 μg/l parameter for pesticides. We have already discussed above that article 9, par. 3 of the Directive provides that the member states cannot derogate from the toxic or microbiological factors, nor is a derogation allowed when it would constitute a public health hazard.

The admissibility of derogations from Directive 80/778 has been judged by the European Court of Justice, with respect to the Italian regulations under discussion.[96] A Piemonte health unit forwarded to the Pretura unificata di Torino (Turin Magistrates' Court) a report containing findings resulting from the analysis of water intended for human consumption, which revealed the presence in a number of wells of atrazine in excess of the 0,1 μg/l limit. The Pretore considered whether the derogations introduced by the orders discussed above came within the limits of the derogation which were allowed under the Directive. He considered further that the lawfulness of these Piemonte orders raised a question of the interpretation of community law. Consequently the Pretore found it necessary to refer to the Court of Justice for a preliminary ruling.

The court held that the derogation provided for in article 9 of the Directive does not concern the facts at issue in the main proceedings. The derogations concerned toxic

94. See Premazzi, G. and Ziglio, G., 1991, p. 10.
95. *Official Journal*, L 230 of 19 August 1991.
96. Case 228/87 of 22 september 1988, Pretura unificata di Torino.

substances and are therefore excluded by article 9(3). The relevant question is whether the Italian derogations could be admissible under article 10(1) of Directive 80/778, which provides:

"In the event of emergencies, the competent national authorities may, for a limited period of time and up to a maximum value to be determined by them, allow the maximum admissible concentration shown in Annex I to be exceeded, provided that this does not constitute an unacceptable risk to public health and provided that the supply of water for human consumption cannot be maintained in any other way".

The court held that these derogations are subject to strict conditions:

- the term "emergencies" must be construed as meaning urgent situations in which the competent authorities are required to cope quickly with difficulties in the supply of drinking water;
- the exceeding of the MAC value may be allowed only for a limited period of time, normally necessary in order to restore the quality of waters effected;
- the excess concentration may not under any circumstances constitute an unacceptable risk to public health;
- there must be no possibility of maintaining the drinking water supply in any other way.

Since this case was a preliminary ruling the Court did not have jurisdiction to rule on the compatibility of a national provision with Community law. It is therefore to the Italian Court to determine whether or not the Italian legislation is compatible with the Directive. It seems rather hard to justify the Italian exceptions under art. 10 (1), as interpreted by the Court of Justice.

This conclusion can be supported by the judgment of the court in the more recent case C 42/89 of 5 July 1990. This case also dealt with a violation of the Directive 80/778, by the Kingdom of Belgium[97]. This case differs from the previous one because it was not a preliminary ruling, but a case directed against Belgium for an alleged lack of implementation of the Directive[98] and because the derogation that Belgium claimed was not based on art. 10, but on art. 9 and art. 20 of the Directive[99]. Belgium had requested additional time under art. 20 in order to comply with the Directive in respect of the water at Verviers. An initial problem was that article 5 of the Decree of the Walloon Regional Executive allowed the Minister to authorize MAC's to be exceeded in the event of serious

97. Case 42/89 of 5 July 1990, Commission of the European Communities v. Kingdom of Belgium, also published in *Milieu en Recht*, 1991, 155-158, with comment by Jans.

98. Therefore in this case the Court had to decide whether Belgian law was compatible with community law.

99. The facts of this case were already discussed in section III.2.

accidental circumstances "in so far as the excess amount does not entail only unacceptable risk to public health and the network distribution cannot be assured in any other manner". The decree in question in no way excluded derogations relating to toxic or microbiological factors, although that is an express requirement of art. 9 (3) of the Directive. Hence, the Walloon decree was found in violation of the Directive in that respect.

The Belgian government requested for a longer period for complying with the requirements of the Directive, as far as the drinking water in Verviers was concerned. This request was denied since it was made more than four years after the expiry of the period for such a request. As regards the difficulties in ensuring that the water supplied to the town of Verviers is in conformity with the Directive, the Court held that:

"a Member State may not plead practical or administrative difficulties in order to justify non-compliance with the obligations and time-limits laid down in Community-Directives. The same holds true of financial difficulties, which it is for the Member States to overcome by adopting appropriate measures".

Hence, the Court declared that by permitting the Walloon Region to allow the MAC's to be exceeded in circumstances other than those provided for in the Directive, and the supply to Verviers of drinking water not in conformity with the requirements laid down in the Directive, Belgium had failed to fulfil its obligation under the EEC Treaty.

Has Italy fulfilled its Community obligations? This question can be analyzed briefly by summarizing the previous sections. The Directive 80/778 forced Italy (and every other member state) to formally implement the Directive by July, 17, 1982 and to meet the parameters of Annex 1 by August 1985. As far as the formal implementation is concerned the least one can say is that Italy has not implemented the Directive before the deadline. A formal implementation took place by DCPM on 8 February 1985 and this was only executed by DPR of 24 May 1988. It is also questionable, given the above discussed EC case law, whether the actual implementation will satisfy the EC Court of Justice. The judgment in case 228/87 as preliminary ruling on the question by the Pretura di Torino made clear that the Piemonte derogations can hardly be upheld under article 10 of the Directive since there were no "urgent situations in which the national authorities are required to cope suddenly with difficulties in the supply of water for human consumption".

Also, the national legislation of 14 February 1989 allowing for a derogation for atrazine up to 0,8 $\mu g/l$ seems problematic under EC Law, since the recent case C 42/89 against Belgium made clear once again that derogations from the MAC values of the Directive are allowed only under the strict conditions laid down in the Directive.

Finally, Italy could incur EC liability not only for a lack of formal (correct) implementation, but also for an actual violation of the drinking water standards, as in the Verviers case. The Milan paper by Vighi et.al. seemed to indicate that pesticides can still be found in drinking water at values higher than 0,1 $\mu g/l$, which is in violation of the

Directive[100]. In this respect one should remember that the output-orientation of Directive 80/778 means that member states are free to chose the mechanisms to reach the goals set. However, if a prohibition of the sale and use of atrazine (as proclaimed in Italy by ordinances of 21 March 1990 and 6 February 1991) does not inhibit other pesticides to reach the drinking water, Italy could still incur EC liability as long as the levels of individual pesticides in drinking water are higher than 0,1 μg/l, be it atrazine or any other pesticide, or higher than 0.5 μg/l in total[101].

3.4. SANCTIONS AND ENFORCEMENT

After having described the various regulatory mechanisms which a Country could choose to implement the Directive, and having discussed the way in which the actual implementation in Italy took place, we should now examine whether these regulations are effectively enforced. We have indicated above that the EC Court of Justice requires a member state to impose sanctions upon the violation of a regulation which has implemented an EC Directive. These should not necessarily be criminal sanctions, but they should at least be "effective and deterrent". Before we will describe the sanctions in the Italian legislation we will briefly discuss the economic theory of criminal law, which gives a useful insight in the function of criminal sanctions.

3.4.1 Some economics of crime and punishment

We have explained in section I why environmental policy in general cannot depend only upon liability rules and why governmental regulation of environmental standards seems necessary. We now turn to the question of how optimal sanctions should be constructed. According to the economic literature, the use of monetary sanctions is justified in cases of relatively high damage and a relatively low chance of actually catching the offender(s)[102]. Indeed, since the probability of being caught when violating a regulatory norm is less than one, a sanction is needed which exceeds the magnitude of the harm. The formula used in this literature for deciding how large a fine (D) should be if the chance (p) that the offender will be actually caught and forced to pay the fine is less than one is:

$$D = L/p \tag{1}$$

100. Also in the paper by the subgroup of economists mentions that in 1989 in Piemonte about 21% of the analyses of drinking water revealed atrazine concentrations exceeding 0,1 μg/l, which violates the Directive (see Section 4.1).

101. The paper by the subgroup of economists mentions that in some areas atrazine will be replaced by linuron and terbuthilazine, which might lead to levels of these compounds in drinking water, exceeding the EC standard.

102. Posner, R., "An economic theory of the criminal law", *Columbia Law Review*, 1985, 1193-1231.

where L is the harm caused by the offender in the case where he is caught. If p = 1, L and D are equal, assuming a risk neutral offender. But if, for example, L = *f* 10,000 and p = 0,1, meaning that nine times out of ten the offender escapes the clutches of the law, then D, the optimal fine, is *f* 100,000. Only then is the expected fine cost to the prospective offender (pD) equal to the harm of his act (L).

Let us now turn to the problem of criminal law and examine the costs of crime as perceived by the offender. A presupposition of the economic analysis of criminal law is that a rational prospective offender can be viewed as a profit maximizer. This offender will therefore take both the costs and the benefits of committing a crime into account and not take illegal action unless the expected benefits of the crime exceed the expected costs[103]. To deter a prospective offender one should therefore increase these expected costs. In calculating the expected costs, two factors are of importance: the probability that the offender is caught and convicted (p) and the expected maximum punishment (S). The multiplication of these factors then constitutes the expected cost of a crime (C), so that

$$C = pS \tag{2}$$

It follows that there are two ways to increase C to the level where they exceed the expected benefits, namely by increasing p or S. If for example S = *f* 100,000 and p = 0.01, then C = *f* 1,000. By raising either the probability of getting caught or the expected maximum punishment with a factor 2, the expected costs in both cases will double (C' = 2C). However, these two ways to obtain this result are not equally easy to achieve. An increase in the expected maximum punishment evidently demands less effort from the government, and consequently will inflict less costs on society, than an increase in (and future stabilisation of) the probability of catching the offender[104].
Hence, there is a trade-off between optimal deterrence and enforcement costs.

An important lesson is that a rational polluter, assuming that he is risk neutral, will pollute if the benefits outweigh his expected costs. Applying these thoughts to the harm caused by the use of atrazine one should take into account both the magnitude of the harm and the expected benefits for the offender. The basic problem with criminal sanctions in general, but especially in the field of environmental crimes, is that the probability of being detected may be quite low and thus also the deterrent effect, unless this can be compensated by extremely high fines. For instance, let us assume that the harm caused by

103. Bowles, R., *Law and the Economy*, 1982, Oxford, Martin Robertson, 57; Shavell, 1985, 1241. See the seminal article by Becker, G., "Crime and Punishment: An Economic Approach", *Journal of Political Economy*, 1968, 169-217.

104. See Polinsky, A.M., and Shavell, S., "A note on optimal fines when wealth varies among individuals", *American Economic Review*, 1991, 618-621.

pesticide X instead of a harmless pesticide[105] is $ 500,000 and the detection rate is 1%. Thus the fine that makes the expected costs for the criminal equal to the harm is $ 50.000.000. This might be impossible to collect. The detection rate could be increased, but that is costly. Therefore the expected costs of committing the crime (using the harmful pesticide) end up as being less than the social harm[106]. Let us now turn again to the situation of the prospective risk neutral offender. If the detection rate were again 1% and the fine were e.g. $ 1 million, the expected costs of using the forbidden pesticide would be $ 10.000. If these costs would outweigh the benefits of using the prohibited pesticide compared to substitutes, the crime might still be committed.

This illustrates a general problem in the enforcement of environmental regulations through the use of the criminal law in that sanctions are often relatively low, compared to the social harm done; judges often do not see these offences as "serious" crimes and so impose relatively small fines and though a lack of expertise and manpower expertise is often low, which leads to law enforcement being often combined with other enforcement mechanisms such as administrative sanctions. Moreover, if the assumption of risk neutrality is relaxed, one can understand why even relatively small fines might have a deterrent effect: a risk-averse individual might even be deterred by smaller fines. Finally, the control and effectiveness of the criminal law can often be increased by formulating the prohibited behaviour in the legislation in an effective way. One example would be not only to prohibit the use of a harmful pesticide (which is hard to detect), but also to prohibit the importation and sale of a pesticide in a country. It is, of course, much easier to control a few firms that produce chemicals than to control every farmer that might use the harmful pesticides.

3.4.2 The crimes and sanctions in the Italian regulation

The only Italian regulation in which we found explicit sanctions is the DPR of May 24, 1988, implementing EEC Directive 80/778. This is a typical output-regulation in which the sanctions are therefore mostly directed towards the waterworks and not towards the emitter of toxic substances. The ordinances of 21 March 1990 and 6 February 1991 prohibiting the sale and use of atrazine in Italy do not have any provision concerning sanctions. This does, however, not mean that no sanctions apply at all. Most likely these ordinances have a statutory basis in an Act that might provide that the regulations taken in execution of that Act are punished with criminal sanctions. This will have to be examined more closely.

105. The comparison is of course difficult because the less harmful pesticide might also be weaker from an agronomic viewpoint, so that its use still results in a welfare loss. We assume, however, that the legislator reached an efficient decision in prohibiting pesticide X and in only allowing other harmless pesticides. The only question therefore is how an effective enforcement of this decision can be achieved.

106. See Faure, M. and Heine, G., "The Insurance of Fines: the Case of oil Pollution", *Geneva Papers on Risk and Insurance*, 1991, 47-49.

Moreover, it might well be that as well as to the sanctions in the Italian regulations which have been discussed in this paper, other sanctions are applicable. Most likely Italy has provisions in its criminal code punishing those who pollute the groundwater or the surface water or there might be specific legislation with the same goal[107]. Not having any information in that respect yet, we will focus mainly on the sanctions in DPR Nr. 236 of 24 May 1988.

Article 21 of this Decree holds that everyone who violates the provisions of this decree and supplies water for human consumption that does not fit the quality requirements provided for in Annex I, is punished by a fine ("ammenda") of 250.000 Lire up to 2.000.000 Lire or with an "arresto" (imprisonment) of a maximum of 3 years. According to art. 21 (2) the same punishment applies to those who use water that does not meet the quality requirements of Annex I in food products and to those who in general bring products or substances which are destinated for human consumption on the market, if the water used can cause a health risk. A higher fine of 500.000 to 5.000.000 Lire applies in case of a violation with respect to especially protected areas.

It seems that the formulation of the crimes in this provision is directed, not against those who cause the pollution, but against those who provide a water supply to the public, directly or indirectly, that does not meet the requirements of the Directive, as incorporated by the DPR of 24 May 1988. In most of the cases this will be the water companies. Since the offender will be generally not just an individual but a corporation, the question arises whether Italian criminal law accepts the criminal liability of a legal entity[108]. Italy still sticks to the classic doctrine that only an individual can incur criminal liability. Therefore, if an offence is committed by a corporation, the judge will have to decide which individual should incur liability for that act. Such a person might often be difficult to trace. This might also explain why in Case 228/87 the Pretore of Torino had instituted criminal proceedings against persons unknown. Apparently the Pretore considered that there was liability of an administrative agency since criminal proceedings were instituted against persons unknown for the failure to carry out official duties, an offence under art. 328 of the Italian Criminal Code[109].

A difficult question to answer is whether the sanctions we have been describing can be expected to have a sufficient deterrent effect. It is well known in the literature on environmental law that a prison sanction is almost never imposed for environmental crimes. The fines of 2.000.000 Lire are considered to be very low[110].

107. See e.g. the so-called Legge Merli (legge sull' inquinamento delle acque) of 10 May 1976, Nr. 319 (Heine, G. and Catenacci, M., "Umweltstrafrecht in Italien", in *Zeitschrift für die gesamte Strafrechtswissenschaften*, 1989, 167.

108. As is the case in the Netherlands (art. 51 of the Dutch Criminal Code).

109. Hence, this was in fact a case of "Amtsträgerstrafbarkeit": liability of government officials for environmental crimes (see Faure, M., Oudijk, J.C. and Koopmans, I., 1992).

110. Heine, G. and Catenacci, M., 1989, 172-173.

Moreover one can expect that the costs of removing pesticides from the groundwater are generally substantially higher than this rather marginal fine. There is of course also no 100% detection rate. So given the low fines and low detection rates one can only conclude that these criminal sanctions alone will not give the waterworks appropriate incentives for investments in water treatment to achieve the values set in Directive 80/778. If the waterworks do strive for these values there might either be high risk aversion with the individual civil servant for a criminal prosecution or other mechanisms (e.g. reputation of the waterworks) might play an important role.

It should be highly interesting to examine how the prohibition of the sale and use of atrazine is enforced in Italy and whether there is a relatively high degree of compliance. Especially since farmers might still have "historical" stocks of atrazine, the possibility that these would still be used for quite some time cannot be excluded. More practical information on the effectiveness of the control by administrative authorities with respect to the prohibition of the use of atrazine would be needed to judge, in general, the likely success of the enforcement policy. In this respect we were told[111] that atrazine is still produced in Italy, which is legal since the prohibition only covers the sale and use of atrazine, not the production. However, as far as the sale of atrazine is concerned there would be almost complete compliance with the prohibition. All the atrazine which is produced in Italy is exported and pesticides are only sold through a few large warehouses that would respect the bars. As far as the use is concerned there would be no import of atrazine from countries where the sale is still allowed, but there seems to be a problem with the remaining "historical" stocks of atrazine which can lead to illegal use.

3.5. CONSEQUENCES OF A BAN OF ATRAZINE FOR THE FREE MOVEMENT OF GOODS

Up to now we have mainly examined the implementation of Directive 80/778 and the case law of the EC Court of Justice that developed concerning the implementation of that Directive. A different, though related, question is whether Italy, by merely banning atrazine, has not distorted the competition between the member states or endangered the free movement of goods.

Suppose that Italy was the only member state that would have decided to implement Directive 80/778 by imposing a complete ban on the sale and use of atrazine[112]. Suppose

111. Information by Marco Vighi.

112. As far as we know atrazine use is still allowed in many member states (at least in Belgium, France, Germany, the United Kingdom and the Netherlands); also the Premazzi/Ziglio paper seems to indicate that not many other countries have opted for a complete ban of atrazine. This might change after the implementation of the Directive of July 15, 1991, concerning the placing of plant protection products on the market. This Directive should remove trade barriers by harmonizing registration procedures for pesticides. In Sweden, however, the Swedish National Chemicals Inspectorate decided on May 16, 1989,

further that atrazine has much better agronomic qualities than any other substitute. In these circumstances a ban on atrazine in Italy might constitute a distortion of competition. It could indeed lead to higher agriculture prices in Italy, thus putting Italian farmers in a disfavourable position. If this would indeed be the case, an Italian farmer might try to challenge Italy's implementation of Directive 80/778 by banning atrazine. This is especially so because the Considerations preceding the Directive expressly state that the Directive is exactly meant to avoid differences in the conditions of competition[113]. It is nevertheless not worthwhile to pay too much attention to this speculative idea. First of all it is not clear whether this sentence in the consideration has binding force for the member states; in addition there is no literature indicating that such a ban of atrazine might violate the goal of the Directive. Moreover, as soon as the Directive on the marketing of plant protection products is implemented, the possible distortions of competition should be avoided by harmonised registration schemes.

There is, however, some case law with respect to a related question of whether a ban of atrazine is a violation of the free movement of goods within the community, as guaranteed by art. 30-36 of the EEC Treaty. If indeed only one member state decides that the sale and use of one product on its territory are prohibited, such a decision inhibits a free movement of that good. The relevant question is of course whether this restraint on the free movement of goods is allowed according to the EEC Treaty.

States are, as a result of the EEC Treaty, not entirely free any more in their laying down of domestic environmental legislation. They have to take into account the relevant obligations under the Treaty. With respect to the free movement of goods the principle of mutual recognition operates. According to this principle rules from another member state that gives an equal protection to the interest invoked (e.g. public health), as the national rule, must be recognized as being equivalent. Therefore the argument that goods (or persons) from another member state are not conform to the national legislation cannot be upheld if those goods are in accordance with equivalent legislation of the other member states.

Article 30 of the EEC Treaty therefore provides that a member state can, in principle, not prohibit the import of a product from another member state. Already in the Dassonville-case of 11 July 1974 the Court of Justice made it clear that any measure that can disturb interstate commerce, be it directly or indirectly, comes into article 30 EEC Treaty[114]. From this rule it would follow that also products that are harmful to the environment would enjoy the free movement of goods. An exception to that rule is provided in article 36: restrictions to the free movement are allowed if necessary for the protection of human health, plants or animals. These restrictive measures may, however,

to cancel the registration of all pesticides containing atrazine.

113. See section II, 3, a, footnote 31.

114. Court of Justice Case 8/74 of 11 July 1974.

not be applied in a discriminatory way. The measure should also be necessary given the goal that is to be reached[115]. An important question was whether an import of certain goods could be restricted only on environmental grounds if there is no health danger. This was generally denied in the literature[116], but in the Cassis de Dyon-case the Court of Justice allowed those restrictive measures, provided that they are applied in a non-discriminatory way; i.e. they should be applicable both to national and to imported products. Also in other cases the Court of Justice accepted that under strict conditions restrictive measures can benefit the free movement of goods if justified by an "environmental" goal[117].

This case law concerned a situation where the import of atrazine would be prohibited, which is, as far as we can see not yet the case. The prohibition of sale and use have of course the same restrictive effect and could only be allowed if it is applied in a non-discriminatory way, which seems indeed to be the case. There exists also case law with respect to a related problem, being the question whether a member state can require additional testing before a product can be put into the market, when it has already undergone testing in another member state. Such extra tests are often costly and could therefore act as trade barriers and hinder the free movement of goods. As we mentioned above, the principle is the mutual recognition. Here again the question arises whether an exception applies for products with a safety or health risk, such as pharmaceutical products or pesticides. It does not seem directly relevant any more for the Italian atrazine case, since no further testing or registration is required before the product can be marketed, but atrazine is banned altogether. The case law concerning registration procedures for pesticides is nevertheless interesting for the other pesticides and it gives a good view of the Community policy in this respect.

By judgment of 17 December 1981 the Court of Justice had to decide whether the Dutch regulation on registration of pesticides was compatible with Community law[118]. A company had marketed a pesticide Fumicot, which was allowed on the market in France, but not in the Netherlands. The company was prosecuted in the Netherlands for having violated the Dutch regulation, since the pesticide was not registered in the Netherlands. The Dutch Criminal Court asked a preliminary ruling to the Court of Justice to know whether the Dutch regulation violated community law. The Court held that a member state can ask for a new registration and for new testing for a product which is already on the market in another member state, provided that they do not require unnecessary extra tests if these are already available from the other registration procedure. The conclusion is therefore that it is not prohibited by article 30-36 EEC Treaty for a

115. Court of Justice Case 104/75 of 20 May 1976, De Peyper.
116. See Jans, J., "Europeesrechtelijke grenzen aan nationaal milieubeleid", *S.E.W.*, 1989, 218.
117. EC Court of Justice Case 302/86 of 20 September 1988. See also Jans, J., 1989, 221.
118. Case 272/80, Court of Justice, 17 December 1981, *S.E.W.*, 1982, 721-726. For a comment see Vogelezang-Stoute, E.M. and Matser, E.J., 1990, 52.

member state to ask for an *ex ante* registration of pesticides, even if these pesticides are already admitted in another member state, provided that no unnecessary testing is required, if testing has already been done in the other state and the results are made available[119].

Part 4
General Conclusions

In this paper we addressed several of the relevant problems, from an institutional perspective, associated with the standard for pesticides in the EC Directive on drinking water. In part 1 of the paper we started with a very basic question, which was how regulation should intervene to prevent the occurance of environmental damage. In this part attention was especially given to the difference between *ex ante* regulation through safety regulation and *ex post* liability rules. It was also shown that in many cases the regulatory process in practice does not follow this ideal model, because of the influence of interest groups which succeed in reducing the efficiency of legislation as a result of lobbying.

This model is further used in an institutional analysis of the drinking water Directive 80/778. First, the question is addressed of whether a regulation of pesticides is in any way necessary. This question is answered affirmatively, since the information on the potential harm that a pesticide could cause is more readily available to government than to the individual user of the pesticide. Moreover the chance that a liability suit would be brought by a victim of pesticide usage is relatively low. Hence, liability rules will only have a minor deterrent effect. We then turned to the contents and structure of the directive 80/778 and gave an overview of the considerations, the text and the annexes. The basic question for the present research was of course why the parameter for pesticides had been set at 0.1 μg/l. It is striking that no answer in that respect could be found, either in the Directive itself, or in its legal history. Apparently the standard was set at the minimum concentration of certain pesticides that could be detected by the analytical methods available at that time, rather than derived from concentrations associated with harm to man.

The third part of the paper addressed the implementation and enforcement of the drinking water Directive, both in general and then with a particular focus on the situation in Italy. In this respect we first discussed the duty of a state to implement the drinking water Directive. We also discussed case 42/89 of 5 July 1990 by which the European Court of Justice held that Belgium violated the EC drinking water Directive because the drinking water of the city of Verviers did not meet the parameters set in the Directive. Hence, the EC Court of Justice does not accept a mere formal implementation, but has to

119. These principles are often applied in national courts as well, when prosecution for marketing without registration takes place. See e.g. Decision of the Dutch Hoge Raad of 20 November 1990, *Milieu en Recht*, 1992, 243-245.

be satisfied that the standards set in the Directives are met by the member states. This can lead to serious consequences for the member states, especially since the new Treaty of Maastricht allows the Commission to ask the Court of Justice to condemn a member state which has not fulfilled its community-obligations to make a penalty payment.

Regarding the implementation of the drinking water Directive we then discussed the comparative benefits of input versus output monitoring. We especially addressed the differences in the monitoring requirements between the two instruments. In case of output-control the monitoring (and therefore the sanctions) are especially directed towards the water company delivering the drinking water, whereas the input-control focuses on the user of the pesticides. We therefore discussed the relationship between the drinking water Directive and the registration of pesticides. We then turned to the implementation of the Directive by Italy and gave a chronological overview of all the Italian regulations regarding drinking water. We then addressed the question of whether Italy fulfilled its community obligations and concluded that the derogations are unlikely to satisfy the EC Court of Justice. Moreover, Italy could incur EC liability not only for a lack of formal implementation, but also for an actual violation of the drinking water standard. The main problem is that the Italian regulation seems to focus on atrazine in particular. But Italy can still incur EC liability if the levels of other individual pesticides in drinking water are higher than the standards set in the drinking water Directive.

We also discovered that, as is the case in most environmental regulations, the sanctions in the Italian legislation are relatively minor and thus will not provide appropriate incentives for preventing pollution. Nevertheless the ban on atrazine seems to have had some effect on its usage. Atrazine is sold only through a few large suppliers who would comply with the ban. However, there is still a problem with the remaining "historical" stocks of atrazine, which could lead to some illegal use.

Finally we looked at the consequences of the Italian ban on the use of atrazine in the context of the free movement of goods. First, we discussed whether Italy, by merely banning atrazine, has distorted the competition between the member states. In addition, we discussed whether the ban on atrazine violates the free movement of goods which is guaranteed under the EC Treaty. EC case-law seems to allow restrictions to be placed on the free movement of goods, if these restrictions are necessary for the protection of human health, plants or animals, provided they are not applied in a discriminatory way.

Looking at points for further research, one should mention that a more careful institutional analysis could be devoted to the various pesticides registration schemes and to the new Directive of 15 July 1991 with respect to pesticides registration. More attention in general could also be paid to the relationship between the various EC Directives regulating water quality. The paper showed that there are various Directives setting specific quality requirements e.g. for drinking water or for surface waters used for drinking water and on the other hand Directives regulating the input of dangerous substances into the waters. The relationship between these "immission" Directives and "emissions" could be examined more carefully. The same is true for the difference between input and output control. In this paper we only indicated some differences between the two regulatory approaches, e.g. as far as monitoring is concerned, but it would merit a careful analysis, e.g. a cost-benefit study, to find out what the optimal mode of control of water quality is.

Finally, it would be interesting to obtain more practical information on the enforcement in practice of the relevant legislation in Italy. The sanctions provided to encourage the prohibition of the use or sale of atrazine in Italy seem relatively minor. Nevertheless, there seems to be a relatively high degree of compliance with this ban of atrazine. Therefore it would be interesting to examine the administrative and legal structure in Italy in more detail to examine how the ban on atrazine is actually enforced.

Part 5
References

Ackerman, B., Rose-Ackerman, S., Sawyer, J. and Henderson, D., *The uncertain search for environmental quality*, London, McMillan, 1974, p. 260-281.

Becker, G., "Crime and Punishment: An Economic Approach", *Journal of Political Economy*, 1968, 169-217.

Bocken, H., "Alternatives to Liability and liability insurance for the compensation of pollution damages, *TMA*, 4-87 and *TMA* 1-88.

Bongaerts, J. and Kraemer, A., *The Law and Economics of water works. Meeting the Increasing Cost of Water Production with Deteriorating Quality of Ground Water*, Berlin, September 1989.

Bongaerts, J. and Kraemer, A., "Water pollution charges in three countries", *European Environment Review*, 1987, Vol. I, No. 4, p. 12-19.

Bowles, R., *Law and Economy*, 1982, Oxford, Martin Robertson, 57.

Brinkhorst, L.J., "De uitvoering van EG-milieubeleid in Nederland: het standpunt van de Europese Commissie", *De uitvoering van EG-milieurichtlijnen, toegespitst op milieugevaarlijke stoffen*, Zwolle, Tjeenk Willink, 1989, 8.

Brown, J.P., "Toward an economic theory of liability", *Journal of Legal Studies*, 1973, 323-349.

Buchanan, J., Tollison, R. and Tullock, G., (eds.), *Toward a theory of the rent seeking society*, University of Michigan Press, 1962.

Buchanan, J. and Tullock, G., *The calculus of consent*, University of Michigan Press, 1962.

Buchanan, J. and Tullock, G., "Polluters profits and political response: direct controls versus taxes", *American Economic Review*, 1975, 139-147.

Coase, R., "The problem of social cost", *Journal of Law and Economics*, 1960, 1-44.

De Vries, P.J.R. and Wilpshaar, J., "Gebruik van bestrijdingsmiddelen door waterschappen", *Waterschapsbelangen*, 1992, 43-49.

Donkers, R.H., "De uitvoering van EG-milieubeleid: het standpunt van de overheid", *De uitvoering van EG-milieurichtlijnen, toegespitst op milieugevaarlijke stoffen*, Zwolle, Tjeenk Willink, 1989, 19.

Eeckman, P. and Vanhees, H., "Overzicht van de rechtspraak van het Hof van Justitie en van het gerecht van eerste aanleg van de Europese gemeenschappen" (september 1990-juli 1991), *Rechtskundig Weekblad*, 1991-92, 1351-1362.

Faure, M. and Bergh, R. van den, "Negligence, strict liability and regulation of safety under Belgium law: an introductory economic analysis", *Geneva papers on Risk and Insurance*, 1987, 109-111.

Faure, M. and Bergh, R. van den, "Liability for nuclear accidents in Belgium from an interest group perspective", in: *International Review of Law and Economics*, 1990, 10, pp. 241-254 .

Faure, M. and Heine, G., "The Insurance of Fines: the Case of oil Pollution", *Geneva Papers on Risk and Insurance*, 1991, 47-49.

Faure, M., Oudijk, J., and Koopmans, I., "Ökonomische Analyse der Amtsträgerstrafbarkeit. Eine Skizze strafrechtlicher Steuerung von Umweltdelinquenz", *Zeitschrift für Wirtschaft-Steuer-Strafrecht*, 1992, 121-129.

Geelhoed, L., Kuiper, M., Braakman, A., Vlas, P., in *Doorwerking van het Europees Recht in de nationale rechtsorde*, Koninklijke Vermande bv, 1986.

Hahn, R., "Economic prescriptions for environmental problems: how the patient followed the doctor's orders", *Journal of Economic Perspectives*, 1989, 95-114.

Heine, G. and Catenacci, M., "Umweltstrafrecht in Italien", in *Zeitschrift für die gesamte Strafrechtswissenschaften*, 1989, 163-187.

Hirsch Ballin, E., "Notitie betreffende de verhouding gemeenschapsrecht-strafrecht", Tweede kamer, 1991-92, 22300 VI, nr. 39.

Jaffer, Susan M., and Kay, John A., "The regulation of Shop Opening Hours in the United Kingdom", in: *Law and Economics and the Economics of Legal Regulation*, Graf von den Schulenburg, J.M. and Skogh, G. (eds.), Dordrecht, Nijhoff, 1986, 169-183.

Jans, J.H., "Europeesrechtelijke grenzen aan nationaal milieubeleid", *S.E.W.*, 1989, 218.

Keller, R., "Zur strafrechtlichen Verantwortlichkeit des Amtsträgers für fehlerhafte Genehmigungen im Umweltrecht", *Festschrift für Kurt Rebmann zum 65. Geburtstag*, München, Beck, 1989.

Klaus, J., "Umweltökonomie und Umweltpolitik", in *Ökologie, Okonomie und Jurisprudenz*, Wenz, E., Issing, O. and Hofmann, H. (eds.), Munchen, 1987, 10-14.

Kohll, N.E.M., "De uitvoering van EG-milieurichtlijnen, toegespitst op milieugevaarlijke stoffen", *De uitvoering van EG-milieurichtlijnen toegespitst op milieugevaarlijke stoffen*, Zwolle, Tjeenk Willink, 1989, 71-95.

Landes, W. and Posner, R., "Tort law as a regulatory regime for catastrophic personal injuries", *Journal of Legal Studies*, 1984, 417.

Landes, W. and Posner, R., *The economic structure of tort law*, Cambridge, Harvard University Press, 1987, p. 42-53.

Lauwaars, R. and Timmermans, C., *Europees Gemeenschapsrecht in kort bestek*, Wolters-Noordhoff, Groningen 1989, 33.

Mackaay, E., *Economics of information and the law*, Boston, Kluwer, 1982.

Maloney, M. and McCormick, R., "A positive theory of environmental quality regulation", *Journal of Law and Economics*, 1982, 100-108.

Mazey, S. and Richardson, J., "British Pressure Groups in the European Community: The Challenge of Brussels", *Parliamentary Affairs*, 1992, 92-107.

McCormick, R. and Tollison, R., *Politicians, Legislation and the economy: an inquiry into the interest group theory of governments*, 1982.

Olson, M., *The rise and decline of nations*, 1982.

Polinski, A.M., and Shavell, S., "A note on optimal fines when wealth varies among individuals", *American Economic Review*, 1991, 618-621.

Posner, R., "An economic theory of the criminal law", *Columbia Law Review*, 1985, 1193-1231.

Posner, R., *Economic analysis of law*, third edition, Boston, Little, Brown & Co., 1986.

Posner, R., "Theories of economic regulation", *Bell Journal of Economics*, 1974, 345.

Premazzi, G., Chiaudani, G. and Ziglio, G., *Scientific Assessment of EC Standards for Drinking Water Ouality*, EC Commission, 1989.

Premazzi, G. and Ziglio, G., *EEC Regulative Problems: State of the Application of the Directive 80/778*, Paper presented at the Working Group 4 Meeting on 20-22 May, 1991, Milan.

Roberts, K.F., *La Direttiva CEE 80/778 concernente la qualita delle acque destinate al consuma umano: difficolta pratiche e consequenze finanziarie collegate all' applicazione della direttiva*, Mogo, 1989.

Schünemann, B., "Die Strafbarkeit von Amtsträgern im Gewässerstrafrecht", *Wistra*, 1986, 235-246.

Schwartz, A. and Wilde, L., "Intervening in markets on the basis of imperfect information: a legal and economic analysis", *University of Pennsylvania Law Review*, 1979, 630-682.

Sevenster, H., "Criminal Law and EC Law", *Common Market Law Review*, 1992, 29-70.

Sevenster, H., *De gevolgen van de voltooiing van de interne markt voor het nationaal milieubeleid*, Publikatiereeks milieubeheer, nr. 1990/2.

Shavell, S., "A model of the optimal use of liability and safety regulation", *Rand Journal of Economics*, 1984, 271-280.

Shavell, S., "Criminal law and the optimal use of non-monetary sanctions as a deterrent", *Columbia Law Review*, 1985, 1232-1262.

Shavell, S., *Economic analysis of accident law*, Cambridge, Harvard University Press, 1987.

Shavell, S., "Liability for harm versus regulation of safety", *Journal of Legal Studies*, 1984, 357-374.

Shavell, S., "The judgement proof problem", *International Review of Law and Economics*, 1986, 43-58.

Stigler, G., "The economics of information", *Journal of Political Economies*, 1961, p. 213.

Van den Bergh, R., "Le Droit Civile face a l'analyse economique de droit", *Ride*, 1988, 229-254.

Van den Bergh, R., "Belgian public policy towards the retailing trade", in: *Law and Economics and the Economics of Legal Regulation*, Graf von den Schulenburg, J.M. and Skogh, G. (eds.), Dordrecht, Nijhoff, 1986, 185-205.

Vogelezang-Stoute, E.M. en Matser, E.J., *De toelating van bestrijdingsmiddelen. Milieu tussen wet en beleid*, Universiteit van Amsterdam, Centrum voor Milieurecht, 1990.

Vogelezang-Stoute, E.M., "Van Vaponastrip tot Atrazin", *Milieu en Recht*, 1992, 231-242.

Vogelezang-Stoute, E.M. and Simons-Vinckx, M.W.L., "De Bestrijdingsmiddelenwet. Procedures en beoordelingscriteria bij de toelating van bestrijdingsmiddelen", Vereniging voor Milieurecht, 1992/2.

Wheeler, M., "The use of criminal statutes to regulate product safety", *Journal of Legal Studies*, 1984, 593-618.

Williamson, O., "Contract-analysis: the transaction cost approach", in *The Economic Approach to Law*, Burrows, P. and Veljanovski, C. (eds.), London, Butterworths, 1981, 39-60.

Williamson, O., "Transaction costs economics: the government of contractual relations", *Journal of Law and Economics*, 1979, 233-261.

Winkelbauer, W., "Die strafrechtliche Verantwortung von Amtsträgern im Umweltstrafrecht", *NStZ*, 1986, 149-153.

Wittman, D., "Prior Regulation versus Post Liability: the choice between input and output monitoring", *Journal of Legal Studies*, 1977, 193-211.

4. EEC REGULATORY PROBLEMS: STATE OF APPLICATION OF DIRECTIVE 80/778

GUIDO PREMAZZI
Environment Institute
JRC
Ispra

GIULIANO ZIGLIO
Hygiene Institute
University of Milano

1. Introduction

It goes without saying that the purpose of the environmental policy of the European Community (EC) is to improve the environment in Europe. In a national context it is well known that the connection between an item of environmental legislation and a measurable improvement is not always easy to demonstrate and with EC legislation the problem is that much more difficult. Not only is the chain between EC legislation and an actual improvement at least one link longer than in a national context - since EC Directives have to have been translated into national legislation or administrative measures before they can take effect - but a number of different Member States are involved each of which may be affected in different ways. [1]

The analogy of a chain linking legislation to an environmental improvement is in fact not very appropriate in an EC context. A better analogy would be a spider's web with the EC institutions at the centre connected to the Member States at the circumference, and with the Member States connected to one another by strands, some of which pass through the EC institutions but some of which do not. Influences can flow along all these connecting strands, but at this point the analogy breaks down. This is because any movement at one point on the web is not necessarily felt immediately at others along the laterals in the same way as it is along the radials. There is no doubt that the influence of the EC institutions, at the centre, on the Member States, at the circumference, is very much stronger than the bilateral influences between the Member States.

The reason for this is that the influence of the EC institutions stems from their ability to adopt legislation binding upon the Member States, whereas the influences between Member States are usually only the influences that come from example. [1]

The environmental policy of the EC can in large measure be regarded as the sum of the items of this EC environmental legislation together with what is then done to implement

L. Bergman and D.M. Pugh (eds.), Environmental Toxicology, Economics and Institutions, 89-110.

that legislation. Any one item may have been inspired by an existing item of national legislation so the EC legislation may result in very little change in that country. The same item of EC legislation may nevertheless have a significant effect in other countries. Any attempt to assess the achievements of EC environmental policy must therefore involve three steps: *first* an examination of the legislative texts themselves to determine their intentions; *second*, a consideration of how each item of EC legislation has been translated into national legislative and administrative measures in all the countries; and *third*, an assessment of what has happened in practice. [1]

According to Article 189 of the Treaty of Rome a Directive is not, unlike a Regulation, directly applicable in the Member States but merely specifies the result to be achieved leaving "to the national Authorities the choice of form and methods". The Commission therefore does not exercise control over the choice of form and methods: what matters is the result - even if in some respects certain Directives are so precise that they leave the Member States little latitude.

If one analyses the results that are to be achieved by the environmental Directives, one can say that the Community imposes four different kinds of obligation on the Member States [2]:

- the need to adapt national legislation, administrative structures and procedures so that they conform to the rules and procedures set out in the Directive;
- the need to put these rules and procedures into practice;
- the need to ensure that the quality of the environment meets the standards or the ends set out in the Directive; and
- the fourth obligation concerns communications that the Member States must make to the Commission to enable the Commission to exercise control over the other three obligations.

There are therefore two categories of failure to comply with Directives, or in other words to implement them. First, those which can be called *formal* and concern legal relations in the Community, including the primacy of Community law over national law which must conform to it, and the obligation of national Authorities to give an account to the Community institutions. Second, those which concern *practical* implementation of Community rules and which require the achievement of practical results (quality of water, the putting in place of authorisation procedures, monitoring, designations, drawing up plans and programmes, etc.).

As regards the adaption of national legislative and administrative measures to conform with EEC Directives, the practice in the different Member States shows that several approaches have been used including the introduction of a new law in Parliament, regulations made under existing laws, simple administrative circulars, and even reliance on the status quo. Besides the choice of instrument there is the question of conformity itself, that is to say the extent to which Community obligations have indeed been transcribed into national measures and the further question of the control of this conformity.

In all the national legal system the most binding instrument is a law or act of the Parliament. A law will usually state the principles and esssential rules but may also authorise the Government to specify the details of these rules so that they can be put into effect by the responsible Authorities and those on whom obligations are placed. It will itself state, or give the Government the power to state, which Authorities are to be responsible. The form of the instrument by which a Government exercises the power given it by a law varies, e.g. decree, arrété, regulation, statutory instrument, command. If the administrative Authorities already possess the necessary powers to achieve the results set out in a Directive it is only necessary to specify these results and give advice where the obligations are not precise, and this can be done by a simple administrative circular.

This paper examines the specific case of the drinking water Directive 80/778.

2. Brief Summary of the Directive

On July 22, the Commission submitted a draft Directive relating to the quality of water for human consumption to the Council of the European Communities. The objective of this regulation is the protection of public health by establishing standards to be respected by the Member States. They concern the quality of drinking water supplied by public main systems, water stored or delivered in bottles or other containers (but not mineral and medicinal waters) and water used for the production of foodstuffs and drinks.

This Directive was adopted by the Council of Ministers of the European Communities on July 15, 1980 (OJ. 229, 30/8/80). This adoption was considered as a very important step since agreement has been reached between 9 Member States on 62 parameters, their numerical value and their monitoring.

The water covered by the Directive is defined to include all water intended for human consumption whether in its original state or after treatment and whether supplied for consumption or used in food or drink manufacture. Some exceptions can be made for water used in food or drink manufacture. Natural mineral waters (such waters are defined in the EEC Directive 80/777) and medicinal waters are excluded from the Directive, but other bottled waters for human consumption are included.

The Directive has three Annexes. Annex I lists 62 parameters and a figure indicating a Guide Level (GL) and a less stringent Maximum Admissible Concentration (MAC) appears against many of the parameters. When no figure appears, the Annex may nevertheless include comments which have to be observed. The parameters are classified in five classes:

1) organoleptic parameters (colour, turbidity, odour, taste) (Table A);
2) physico-chemical parameters (temperature, pH, conductivity, chloride sulphates, silica, calcium, magnesium, sodium, potassium, etc.) (Table B);
3) parameters concerning substances undesirable in excessive amounts (24 metallic and non-metallic ions and organic compounds) (Table C);
4) parameters concerning toxic substances (13 metallic and non-metallic ions and organic compounds) (Table D);
5) microbiological parameters (Coliforms, Streptococci, Clostridium, Salmonella, Pathogenic staphylococci, etc.) (Table E).

Annex II is concerned with monitoring and lists patterns and frequency of standard analyses. As the population served increases the frequency of monitoring increases. There is an order of priority among the parameters determining the frequency of sampling and analysis. This order of priority is taken into account in the system of standard analysis already in use in certain Community countries. Four types of analysis of increasing complexity are selected (Tables A and B):

- minimum monitoring;
- current monitoring;
- periodic monitoring;
- occasional monitoring in special situations or in case of accidents.

The frequence of these standard analyses is determined by the following factors:

- the capacity of the installations (volume of water produced and distributed in m^3/day);
- the size of population concerned (assuming 200 L/day per person).

Annex III lists Reference methods of analysis.

Member States are to fix values for the parameters in Annex I wherever a figure appears and are to ensure that water meets them. For the parameters in Tables A, B, C, D and E the values fixed must be less than or equal to the MAC figures, and the GL figures are to provide a guide. The parameters in Table F do not apply to water used in food production

Member States are to ensure that any substances used in the preparation of water for human consumption do not remain in concentrations higher than the MAC figure in water made available to the user and that they do not constitute a public health hazard.

Member States may derogate in two circumstances:

i) related to the nature of the ground (permanent derogations - Article 9a), and
ii) related to situations arising from exceptional meteorological conditions (temporary derogations - Article 9b).

The Commission is to be informed of derogations. In emergencies, the MAC figures may be exceeded for limited periods if there is no risk to health and water cannot be supplied in any other way. They may also be exceeded, subject to conditions, where a source worse than A3 has to be used.

Member States may also ask the Commission for a delay in meeting the MAC values in exceptional cases but must propose an improvement programme with a timetable (Article 20).

Regular monitoring is to be ensured by the Member States. Water is to be monitored at the point where it is made available to the user. The sampling points are to be determined by the competent national Authorities. Monitoring is to conform with the patterns and frequency set out in Annex II, and as far as practicable Member States are to use the reference methods of analysis set out in Annex III. Where other methods are used they must be comparable.

3. Difficulties and Problems in the Implementation of or in the Compliance with the EEC Directive

The date for formal implementation of the EEC Directive into national legislation was 17 July 1982. Member States were to ensure that by August 1985 drinking water met values for parameters listed in Annex I. However, the deadline for full compliance (both formal and practical) was 1.1.1986 in the case of Greece and Spain, and 1.1.1989 in the case of Portugal. The values are to be fixed by the Member States and must be as stringent as certain MAC figures and more stringent GL limits. At December 1990 correct formal compliance had not yet been fully achieved by a number of Member States.

Table 1 indicates the parameters of the EEC Directive not transposed into national legislations. It must be noted that Greece, Italy and Portugal are the only Member States that have incorporated all 62 parameters of the Directive; on the contrary in Germany up to 24 parameters and 15 in France and Spain have not been transposed. Instead, Table 2 summarizes the additional parameters and specific substances considered in different legislations of Member States.

Nevertheless, the Directive has provided the occasion for national assessments of drinking water quality and for reviews of national policies. A particular effect has been to stimulate discussions on difficulties of application (technical/scientific, legal and regulatory), on specific parameters (nitrates, pesticides, lead, etc.), and on the adequacy of the Directive.

The present situation concerning the application of the Directive in Member States can be summarized as follows [3]:

- In the majority of Member States guideline values have not been taken into account.
- The parameters for which the Directive does not foresee a value remain in general without value in the Member States or are omitted.
- The MACs of the Directive concerning toxic substances are generally adopted in the national laws with the same values except some omissions, or in a few rare cases, somewhat inferior.
- All the other MAC values are integrated in the national laws with limits equal to those of the Directive or, in a few rare cases, inferior to them, however with some omissions depending on the Member State concerned.
- For parameters which cover more than one substance such as parameters 32, 55 and 56, certain Member States single out toxicological limits for some individual products.
- Specific parameters such as nitrate, pesticides, halogenated compounds, lead and potassium appear as a problem in Member States. Of consequence the compliance of MAC values leads to difficulties in application of the Directive. Then, Member States request the Commission for delays and derogations for certain parameters (Table 3).
- The microbiological parameters of the Directive are adopted in all the countries with the exception of parameter No. 62 in Denmark, Germany, Spain, the Netherlands and the United Kingdom. Parameter No. 61 is not considered in French legislation.
- Member States have differently implemented the contents of Annexes II and III concerning surveillance and analytical monitoring (Table 4). [4]

A summary evaluation below accounts for the difficulties in the application of the Directive.

PRACTICAL DIFFICULTIES IN THE ACHIEVEMENT OF THE DIRECTIVE'S REQUIREMENTS

Items	*Problems, needs and uncertainties*
Structure and organisation	• availability and adequacy at the different levels of activity • management capabilities and prioritization • central structure/agency for water management
Scientific and technical aspects	• availability of analytical methods (for some parameters) • identification of single compound (parameters as pesticides and "other organohalogenated compounds") • criteria for selecting limits/MCL • knowledge about existing situation (frequency and modes of contamination, pesticides in common use in the catchment area, point source characteristics, fate/degradation and by-products, etc.). • scientific approach in water management and decision • technology for the control of the contamination (at the point of discharge, at the treatment plant producing drinking water) • reference structures (standardization of analytical procedures, Quality Assurance programmes)

Political and administrative aspects	• regulation of contamination sources (point/diffuse) • application of legislation in the field of water protection • harmonization of all legislation regarding water aspects • zero limit ("no risk approach") versus toxicology based limit • consultation/confrontation with other parties: water undertakings, consumer associations • conflicts of interests (use of agricultural pesticides against quality of water resources)
Economic implications	• cost evaluation for the implementation of the EEC Directive • cost-benefit analysis • availability of financial resources

4. The Specific Case of Pesticides

In the context of the Directive the term "pesticides and related products" includes a whole range of substances including herbicides, insecticides, acaricides, nematocides and fungicides, as well as PCBs and PCTs. MACs for individual pesticides of 0.1 ug/litre and for total pesticides of 0.5 ug/litre are set.

As pesticides are classified in the Directive as toxic substances, derogations under Article 9 are not available. The Directive provides that in situations where the MAC for pesticides is exceeded, an extension of time to comply with the MAC and a programme of action under Article 20 would need to be agreed by the European Commission.

This parameter covers several hundred different chemical substances of widely differing toxicity, persistence and degradability, many of which were introduced after the Directive was drafted.

The 0.5 ug/litre MAC value was set at the limit of detection for chlorinated pesticides such as lindane and parathion. It may have been seen as a surrogate for a zero standard or as a point from which no deterioration should be allowed. Unlike the MACs for other toxic parameters in the Directive, it was not set on the basis of toxicology. MACs were originally intended to be based on a 85 percentile compliance rather than 100%. [5]

As regards the compliance of pesticide parameter in Member States, derived by comparing national legislations, the present situation is summarized in Table 5. It also indicates the situation existing in other Countries outside the Community. The positions of Member States on this EEC limit are generally favourable, even if United Kingdom has formally requested a revision of this parameter and the Federal Republic of Germany has suspended the limit of 0.1 ug/litre for phytoproducts intended for plant protection. Italy establishes derogations for 4 substances, while France fixes toxicological limits for 3 products. About use restrictions for pesticides, it must be noted that Germany, Italy and the Netherlands have adopted more or less stringent limitations in some water catchment areas.

As regards non-Member States, toxicological limits for each compound are fixed with the exception of Switzerland which has adopted the EEC limits (Table 6).

4.1. UNITED KINGDOM

The United Kingdom has formally requested a revision of this parameter so that limits more closely related to health risks can be used. The UK Government considers the MACs for total and individual pesticides inappropriate for the following reasons:

a) it is not possible to assess whether the MAC for total or individual pesticides has been exceeded, because current analytical methods are unable to detect many pesticides even at concentrations above the MAC;
b) no account is taken of the widely differing toxicities of individual pesticides in applying the same value to each pesticide irrespective of its toxicity.

Information from water undertakings during the last years (as from 1984) on pesticides detected in water supplies have led to the following general conclusions [6]:

a) triazines, particularly atrazine, have been detected in many supplies, derived from both surface and groundwater sources;
b) phenoxyalkanoic acid compounds, particularly mecoprop and MCPA but also 2,4-D and MCPB, have been detected in surface and groundwater derived supplies in intensive arable agricultural areas;
c) other pesticides such as the persistent organochlorine compounds, have been detected occasionally but they were probably associated with pollution incidents (e.g. spillage) rather than regular use;
d) water undertakings have not attempted to monitor those pesticides which might be present in water supplies at very low concentration and for which reliable and sufficiently sensitive methods of analysis are not available.

Table 7 gives the concentrations of those pesticides which have been found regularly in some drinking water supplies, including groundwater resources, at concentrations exceeding the EEC Directive limit for individual substances.

Medical advice is that none of the pesticides so far detected in water supplies exceeding the Directive limit constitutes a health risk at the concentration found.

4.2. ITALY

At the request of the Government of Italy, two consultations (11-13 February 1987, 13-18 July 1987) were organized by the WHO Regional Office for Europe in association with the Istituto Superiore di Sanità in Rome to develop recommendations on guideline levels in drinking water for the eleven herbicides most commonly used in Italy, i.e. alachlor, bentazon, 2- (4-chloro-2-methylphenoxy) acetic acid (MCPA), metolachlor, pendimethalin, propanil, pyridate, simazine and trifluralin.

In setting guideline levels, 10% of the total ADI of these substances was allocated to drinking water. This allocation of dietary intake to drinking water is higher than for many of the substances included in the Guidelines for Drinking Water Quality because residues of the herbicides reviewed are unlikely to be present to any significant extent in food at the time of consumption.

Regarding concentrations of atrazine, simazine, bentazon and molinate measured in Italian regions, where rice and maize are produced (Lombardia, Veneto, Emilia-Romagna, Marche, Piemonte and Friuli Venezia Giulia), the following must be stressed:

a) Among 318 sites examined in total in the above mentioned 6 regions, in most of the cases (93.4%) a single contaminant is present, namely:
   - atrazine 88%
   - simazine 1%
   - bentazon 4.4%

   In particular, atrazine is present in 279 sites with the following distribution:
   - in 49% at a level inferior or equal to 0.1 ug/litre;
   - only in 8% is it present at levels from 0.3 to 0.6 ug/litre, with a few isolated peaks of the order 0.7-0.8 ug/litre.

b) In only 21 of 318 sites (6.6%) at the most two herbicides were present simultaneously, namely:
   - atrazine with simazine: 2.8% (9 sites);
   - atrazine with molinate: 0.3% (1 site);
   - atrazine with bentazon: 3.4% (11 sites).

With the "Decree" of 14 February 1989, Italy has established that for a maximum period of 24 months, the CMA for atrazine, simazine, bentazon and molinate will be 0.8, 0.4, 4 and 0.3 ug/litre, respectively. In the case where more herbicides are present simultaneously, the CMA value is determined with the following formula:

$$(C1/MAC1) + (C2/MAC2) + \dots \leq 1$$

where C1 is the concentration of the first contaminant, C2 the concentration of the second contaminant, and MAC1, MAC2 denotes the maximum admissible values for the single contaminants, respectively.

4.3. FEDERAL REPUBLIC OF GERMANY

The most comprehensive compilation of the occurrence of active ingredients of pesticides in drinking water in Germany is published about every 3 months by the Federal Association of the German Gas and Water Industry (BGW). [3] Over 40 active ingredients have been identified up to now in the ug/litre range (Table 8).

Atrazine seems to be a problem in many federal countries; it has been found in Bavaria in 33% of all results and in Baden-Wüttemberg already 1 of 16 wells examined exhibits contents far above the limit of 0.1 ug/litre; 1 out of 7 wells is contamined at levels between 0.05 and 0.1 ug/litre.

The extensive Drinking Water Monitoring Programme conducted during 1984 till 1986 [7] in several Lander has shown the occurrence of pesticide residues at levels above the EC limit (Table 9). However, many substances may have escaped detection because even for those classified as W1 and W2, the analytical methods for the ug/litre range are not always available (Table 10). From a total of 16 active subsances under the most stringent W1 class, only 7 are accessible for detection in this range (W1). Similar figures exist for the less severe W2 substances of which about 60% cannot be detected due to methodological deficiencies (Table 11).

W1 class include pesticides with the most stringent restricted use; W2 class include less severe substances with restricted use. At present in Germany, 35 substances are banned, 10 are partially banned, a further 23 are banned in all protection zones and another 50 must not be used in inner zones of water catchment areas [8].

4.4. THE NETHERLANDS

The total amount of pesticides which is applied yearly to cultivated soil in the Netherlands is approximately 20 million kg of active chemicals. Were this amount of pesticides evenly spread on all cultivated soil the average application would be 10 kg/ha. For example, in the United States, the states with the highest use, show an average use of 3-4 kg/ha.

In 1985 the National Institute of Public Health and Environmental Protection started a research programme on 14 pesticides in groundwater in four areas with soil types vulnerable to leaching. Though detection limits of the studied pesticides ranged from 0.2-10 ug/litre, four pesticides showed concentrations above the drinking water standard of 0.1 ug/litre, viz. atrazine (max 0.6 ug/litre), dinoseb (max 9.2 ug/litre), 1,3-dichloropropene (max 80 ug/litre) and aldicarb (max 130 ug/litre). The latter two pesticides are already prohibited in groundwater protection areas under the Pesticides Act, which is justified by these results.

The Soil Protection Act (effective from January 1, 1987) forms the legal framework for soil and groundwater protection in the Netherlands. Two levels of protection are distinguished, i.e. the general protection level and a specific protection level. The general protection level will be realized by rules set by the national government. These rules may consist of effect-oriented (setting of quality standards) as well as source-oriented measures towards pollution sources.

The Dutch soil and groundwater protection policy aims at maintaining the multifunctionality of the soil. This implies a quality level of soil and groundwater which allows for the soil to function as a growing place for agricultural crops and as a reservoir for drinking water abstraction, both now and in the future.

Diffuse sources of pollution, e.g. from agriculture, will have to meet the following criteria:

- reaching an equilibrium between input and crop extraction as much as is possible;
- the input of man-made chemicals has to meet requirements on persistency and leachability;
- the concentration of chemical substances in soil and groundwater will have to meet quality standards.

Since 1974 the protection of areas around pumping stations for public water supply was seen as a governmental task in a number of provinces. In 1980 the VEWIN published guidelines for the dimensioning of groundwater protection areas which have been accepted by the national government in 1984. Protection areas are based on the travel time concept with the outer boundary at the 25 year flow distance from the pumping stations. This implies that only a part of the recharge area will receive additional protection based on provincial regulations. At January 1, 1989 all 12 provinces are obliged to draw up a groundwater protection plan and a protecion ordinance under the Soil Protection Act.

The policy against pesticide pollution requires an authorization at the level of the national government. This holds too for restricted authorization concerning the use of pesticides in water protection areas. In 1987 about 10% of the 314 pesticides (active chemicals) which may be used in agriculture are probibited in these areas. Moreover, for another 17 pesticides the use in these areas is restricted with respect to time of application and soil composition.

In fact, since 1974 the restrictions in use were advised to the provincial authorities to be inserted in provincial ordinances. However, enforcement of these regulations left much to be desired. Besides, in a number of provinces these ordinances did not even exist. Since 1982 the Pesticides Act regulates the use of pesticides in protection areas. From 1983 the prohibitions as well as the restrictions are mentioned on the label attached to the package. A blacklist of pesticides not admitted in groundwater protection areas is given in Table 12.

Groundwater quality problems arose especially from the pesticides authorized before 1975. These older pesticides still take a dominant position on the Dutch market.

In order to minimize the groundwater pollution potential of pesticides the following measures will be taken in the Netherlands:

- critical reassessment of the "older" pesticides to decide whether authorization for the Dutch market can be continued;
- the number of pesticides which are not permitted for use in groundwater protection areas will be raised;
- the behaviour of pesticides in soil types vulnerable to leaching will be given more weight in the authorization procedure;
- periodic testing of application devices will be obliged.

The current levels of pesticides in drinking water resources in the Netherlands are summarized in Tables 13 and 14.

5. Conclusions

In a short period of years the Community has introduced a large body of Directives in the field of water. Undoubtedly these have contributed to national developments and have helped to focus attention on national deficiencies. They have therefore helped to raise awareness of environmental issues. However the administrative solutions adopted for particular problems in one country are still little known in others and, despite the Directives, there is still a great diversity in the different Member States. This clearly points to the need for a greater exchange of views and information between countries by the responsible Authorities, at whatever level of government they may be.

An unexpected effect of EEC policy has been to centralise into the hands of national governments some functions that had previously been devolved to regional or local Authorities. The setting of standards at EEC level means that these standards are negotiated by the national Government and therefore can no longer be set entirely at a local level. Furthermore the central Government has to ensure that the standards are applied by local/regional Authorities since it is the central Government which has to report to the EEC Commission that the obligations in the Directive has been fulfilled.

The EEC Directive 80/778 is having important effects in Member States, although its full impact will be easier to assess once the extent of delays and derogations are known. A particular effect has been to stimulate discussions in all Countries on the problem of nitrates, pesticides and lead and the validity of related standards is being called into question.

However, a satisfactory implementation of Directive 80/778 in all Member States cannot be shown. The reason for this is that Member States had not fully realised the difficulties in fulfilling the obligation that they had undertaken. These difficulties involve research aspects, operation and management of water resources as well as legislative administrative procedures. They are set out below and should not however be regarded as a complete catalogue.

Research aspects

- Fate of synthetic compounds
- Experimental and epidemiological data for risk evaluation
- Advancement and reliability of analytical methods
- Integrated approach methodology for the protection and management of water resources.

Operation and management measures

- Economic and financial needs, including prices of water
- Technology for control and treatment and relative costs
- Modelling in water resource protection and management
- Survey and monitoring activity
- Application of polluter-pays principle
- Clear identification of the responsibility
- Coordination between all competent authorities related to the water cycle.

Legislative and administrative procedures

- Scientific and technical revision/updating of the Directive
- Harmonization of water legislations
- Centralized Organisation/Agency in the field of Quality Control and Quality Assurance Programmes.

6. References

[1] Haigh, N., 1986. *European Community Environmental Policy in Practice*, Vol. 1, Graham and Trotman, 107 p.

[21 Collins, K., 1988. Document of the European Parliament, A2-0298/87, 71 p.

[3] Premazzi, G., Chiaudani, G., Ziglio, G., 1989. Scientific Assessment of EC Standards for Drinking Water Quality. EUR Report 12427. Commission of the European Communities, 200 p.

[4] Ziglio, G., Premazzi, G. Scientific Assessment of EC Standards for Drinking Water Quality: Monitoring, Sampling, Frequency and Reference Methods of Analysis (in press).

[5] European Institute for Water, 1989. Seminar on EEC Directive: Practical difficulties and financial implications related to the implementation of the Directive, 16-17 March, 1989. Como.

[6] D o E, 1986. Standing Committee of Analysts.

[7] Industrieverband Agrar e. V., 1988. The IPS Raw Water Monitoring.

[8] Federal Health Office, 1988. Bundesgetzblatt 5702, 4/8/1988, N. 37, Teil 1.

TABLE 1. Parameters not transposed in the national legislations

Belgium	Nine: 5, 6, 10, 16, 18, 19, 30, 45, 54
Denmark	Three: 10, 30, 62
France	Sixteen: 7, 10, 11, 18, 19, 25, 27, 30, 39, 40, 41, 42, 45, 54, 61
Germany*	Twenty-six: 4, 8, 10, 11, 16, 17, 18, 19, 23, 25, 26, 27, 28, 29, 30, 35, 36, 37, 39, 40, 42, 45, 52, 53, 54, 62
Greece	None
Ireland	Eight: 10, 18, 19, 25, 39, 41, 45, 54
Italy	None
Luxembourg	Three: 39, 45, 54
Portugal	None
Spain	Fifteen: 5, 13, 14, 19, 23, 25, 28, 30, 32, 39, 41, 42, 43, 45, 62
The Netherlands	Five: 10, 25, 39, 41, 62
United Kingdom	Ten: 10, 18, 19, 26, 39, 40, 41, 45, 54, 62

* Most of these parameters have been implemented via indirect regulations (e.g. Section 2, subsections 2 and 3 minimization requirement, Trinkwasser Verordnung).

TABLE 2. Additional parameters considered in the national legislations

Denmark	THMs, methane, bicarbonates, fluorescent bacteria
France	Benzo3, 4, pyrene, aldrin/dieldrin, hexachlorobenzene, PCB/PCT
Germany	Surfactants (ionic, nonionic), trichloroethene, dichloromethane, trichloroethylene, tetrachloroethylene, tetrachloromethane, free chlorine, chlorine dioxide, radioactivity
Greece	Radioactivity
Italy	Surfactants (ionic, nonionic), atrazine, bentazon, molinate, simazine
Portugal	Radioactivity
Spain	Radioactivity
United Kingdom	Benzo3, 4 pyrene, tetrachloromethane, trichloroethene, tetrachloroethene, THMs

TABLE 3. Delays and derogations (articles 9 and 20 of the EEC Directive)

Belgium	Applying article 9.1 (a) the Belgian authorities have agreed on 24 derogations for the following parameters: chloride, Na, Fe, Mn, Zn and K. The duration is variable (5-7 years). For K a fixed derogation has been established for a specific area; no demands to the Commission on article 20.
Denmark	Applying article 9.1. (a), derogations for certain installations producing water with a distribution volume of the order of 3000 m^3/year have been agreed upon; the derogations concern nitrates and sodium chloride. This Member Country has not presented demands to the Commission concerning article 20.
Germany	The German authorities, in their national legislation, have established derogations relative to NH_4, K, Mg and sulphates. For pesticides the Commission has been informed about the delay agreed by the authorities in 1986, namely up to 1.10.1989.
France	One derogation under article 9 is established for sulphates and fluorides. As far as nitrates are concerned, the French authorities do not have the intention to establish derogations. Up to now the French authorities have not presented to the Commission any specific demand concerning article 20 of the Directive. They deem a further delay on this article necessary as far as nitrates are concerned.
Italy	The Italian authorities have established derogations relative to NH_4, nitrates, Mg, Mn, Fe, fluorides, sulphates, oxidability, Na, dry residues, colour, odour and taste. For some pesticides (atrazine, simazine, bentazone and molinate) the Commission has been informed about the delay in the observance of EC limits up to 14.2.1991.
Spain	Derogations will be necessary for a certain number of parameters: K, nitrates, Mg, sulphates, Al and fluorides. Up to now the Commission has not been informed about such derogations.
The Netherlands	The Dutch authorities have not yet officialized any request for derogations to be introduced in the various water supply services. They have established a derogation for the parameters phenol and oils. The Dutch authorities have not presented any demand to the commission concerning article 20.
United Kingdom	Applying article 9.1 (a) the British authorities have agreed upon derogations relative to different parameters (Fe, Mn, Al, nitrates) for a certain number of regions. They have submitted to the Commission a particular demand on article 20 concerning a further delay in respect of Annex I regarding Pb, Al and Fe, microbiological parameters, as well as for private resources.

TABLE 4. Summary comparative analysis of procedure adopted in monitoring water quality characteristics (Annexes II and III)

Ch.	GR	DK	GE	UK	IR	I	L	NL	S	P	F	US
1	yes	no	no	no	yes	yes	yes	no	no	no	no	no
2	yes	no	no	no	yes	no	yes	no	no	no	yes	no
3	yes	-	no	no	yes	>	yes	-	-	-	≠	-
4	no	no	yes	yes	no	no	no	yes	yes	yes	yes	yes
5	no	yes	no	yes	no	no	no	yes	yes	no	yes	yes
6	no	yes	yes	yes	no	no	no	yes	yes	yes	no	yes
7	yes	yes	no	yes	yes	yes	yes	yes	yes	yes	yes	no
8	b	vol	vol	b	b	pop	b	pop	pop	pop	b	pop
9	no	no	yes	no	no	yes	no	yes	yes	no	no	yes
10	no	yes	no	no	no	no	no	yes	yes	no	yes	yes

LIST OF CHARACTERISTICS (ch., column 1)

1. Do monitoring programmes and reference methods strictly adhere to what is reported in Annexes II and III?
2. Are the parameters classified according to the EEC Directive?
3. Are the parameters (number type) included in monitoring programmes C_1-C_3 the same of those indicated by the EEC Directive? (>: greater, ≠: different)
4. Are duties and responsibilities clearly indicated?
5. Are there monitoring programmes expressly differentiated by type of source, environmental situation, and water technological cycle?
6. Are criteria listed to define the number of sampling points?
7. Are the sampling frequencies expressly different for surface and groundwaters?
8. Is the frequency calculated on the basis of the volume of water produced/distributed or of the population involved? (vol: volume only, pop: population only, b: both)
9. Is any further technical information included in the analytical procedures?
10. Is there reference to national standardized methods?

TABLE 5. Compliance of Parameter No 55 in Member States

Country	Compliance	National position on EEC limits	Monitoring	Use restrictions	Comments
Belgium	yes (1988)	favourable	programmed	no	EEC limits
Denmark	yes (1988)	favourable	-	no	EEC limits
France	yes (1989)	favourable	running	no	EEC limits toxicological limits for 3 products
Germany	yes (1986)	favourable	running	for 79 products in water supply catchments	EEC limits, in progress definition of toxic limits
Greece	yes (1986)	favourable	-	no	EEC limits
Ireland	yes (1988)	favourable	running by local authorities	no	EEC limits
Italy	yes (1986)	favourable	running by local authorities	yes	EEC limits derogation for 4 substances
Luxembourg	yes (1985)	favourable	-	no	EEC limits
The Netherlands	yes (1984)	favourable	running	yes	EEC limits
United Kingdom	yes (1989)	favourable	running	no	EEC limits, definition of toxicol limits

TABLE 6. Present situation in some non-Member States, including WHO

Country	To be in force	National position on EEC limits	Monitoring	Use restrictions	Comments
Australia	1987		done	no	toxicological limits for each product
Canada	1987		done	no	toxicological limits for each product
Norway	1987			yes	limits for total pesticides
Sweden	1988		running	50% limitation in 1990	-
Switzerland	1985	favourable	no	no	EEC limits
USA	1988		done	no	toxicological limits for each product
WHO	1984				toxicological limit for each product

TABLE 7. Pesticide concentrations in UK drinking water

Substance	**Typical Concentrations** (ug/litre)
Atrazine	0.1 - 2.0
Carbophenothion	0.1 - 0.2
Dimethoate	0.1 - 0.2
MCPA / MCPB	0.1 - 2.0
Mecoprop	0.1 - 2.0
Simazine	0.1 - 0.4
2,4 - D	0.1 - 0.5

TABLE 8. Pesticides found in ground water (GW), bank filtrate (U), wells (B), sources (Q) or drinking water (TW) in West Germany

Restricted use in water catchment areas	**Use in water catchment areas not restricted**
Atrazine (GW, TW, B, U)	Azinophos-ethyl (U)
Bentazon (B)	Chlorphenvinphos (GW)
Bromacil (B)	Chloridazon (B)
Clopyralid (GW)	Chlortoluron (U, GW, TW)
1,2 - dichloropropane (B)	2,4-D (U, B)
1,3 - dichloropropane (GW)	Desethylatrazine (GW)
Dimethoate (GW, U)	Desethylsimazine (GW)
Dinosebacetate (B)	Disulfoton (U)
Etrimfos (GW)	Diuron (U)
Lindane (U, GW)	Isoproturon (B)
Metazachlor (U, GW, B)	Linuron (GW)
Methylisothiocyanate (B)	MCPA (GW)
Propazine (GW)	MCPP (Q, GW, B)
Pyridate (B)	Methabenzthiazuron (GW)
Simazine (GW, B, U, Q)	Terbutylazine (GW)
2,4, 5-T (GW)	Metobromuron (U, GW)
	Metolachlor (U, GW, B)
	Metoxuron (U)
<u>Not allowed at all</u>	Monuron (U)
	Parathion - equivalents (U)
Aldrin (GW)	Pendimethalin (B)
Heptachlor (GW)	Prometryn (GW)
Hexachlorbenzene (U)	Vinclozolin (B)

TABLE 9. Results of the IPS Raw Water Monitoring Programme (Industrieverband Agrar e. V.)

LÄNDER	No of sampled wells	Analytical findings above the EEC limit		
		No of wells	No of active ingredients detected	Notes
Baden Württemberg	65	6	2	atrazine bentazone
Bavaria	21	5	4	atrazine bentazone chloridazone simazine
Hesse	10	0	0	
Northrhine-Westphalia	45	3	1	bentazone
Lower Saxony	38	1	1	pyridate
Rhineland-Palatine	5	1	2	bentazone CMPP
Schleswig-Holstein	22	4	1	dichloropropane

TABLE 10. Pesticides with Restricted Use in Water Catchment Areas

Active ingredient W1 ban	**Detection method in the ug/litre range** available	not available
Aldicarb	+	
Amitrol	+	
Bromacil	+	
Chlorambene		-
DNOC		-
Dazomet		-
Dikegulac		-
Dichlorpropene	+	
Metam-sodium		-
Methylisothiocyanate	+	
Sodium chlorate		-
Oxamyl		-
Picloram	+	
Tebuthiron		-
Terbacil	+	
Thiofanox		-
TOTAL	**7**	**9**

TABLE 11. Pesticides with Restricted Use in Water Catchment Areas

Active ingredient	**Detection method in the ug/litre range**	
W2 ban	available	not available
Alloxydim-salt		-
Anizalin		-
Afrikal 67		-
Asulam		-
Atrazine	+	
Bac. thring.	+	
Bendiocarb		-
Benodanil		-
Bentazone	+	
Carbetamide		-
Carbofuran	+	
Chinonamide		-
Chlorthiamide		-
Cyanazine	+	
Dicamba		-
Dichlobenil		-
Dimethoate	+	
Dinoseb (-acetate)	+	
Dinoterb	+	
DNOC		-
Ethidimuron		-
Ethiofencarb		-
Ethoprophos	+	
Fenaminosulf		-
Hexazinon		-
Isocarbamid		-
Karbutilate		-
Lindane	+	
Maleic acid hydrazide		-
Meifluidid		-
Metalaxyl	+	
Metazachlor	+	
Methomyl		-
Methylbromide		-
Monochlorbenzene		-
Napropamide		-
Oxycarboxine		-
Pirimicarb		-
Propoxur	+	
Pyridate	+	
Secbumeton		-
Terbumeton		-
2,4,5 - T and salts	+	
TCA		-
Thiazfluron		-
TOTAL	**15**	**31**

TABLE 12. Residues from Pesticides not admitted in Groundwater Protection Areas

Compound	**Concentration** (ug/litre)
1.3-dichloropropene	<0.10 - 80
1.2-dichloropropane (contamination in 1.3-dichloropropene)	<0.10 - 160
1.2.3-trichloropropane	0.10 - 30
1.2.2-trichloropropane	0.10 - 1
aldicarb (incl. metabolites)	4.50 - 130
metolachlor	<0.04 - 0.21
bromacil	0.10 - 2.0
bentazone	<0.10 - 1.0
atrazine	<0.02 - 0.8
desethyl - atrazine	<0.05 - 0.3
desisopropylatrazine	<0.05 - 0.28

TABLE 13. Pesticides frequently found in Surface Water

Contaminant	**Concentration** (ug/litre)
atrazine	0.10 - 1
bentazone	0.10 - 2
dichloropropane	0.10 - 1
simazine	0.10 - 1
metolachlor	0.01 - 0.1

TABLE 14. Pesticides observed in Groundwater

Contaminant	**Concentration** (ug/litre)
ethoprophos	<0.05 - 1.5
dinoseb	<0.05 - 9.2
metamitron	0.06 - 0.73
ethylene - thioureum (melabolite of maneb. zineb. etc.)	<0.50 - 2.0
simazine	<0.15 - 0.5
oxamyl	<0.10 - 0.2
dichlorobenzamide	0.10 - 0.3
cyanazine	0.10 - 1
diuron	0.05
dichloropropane	0.10 - 0.5
bromacil	0.10 - 2.0
methylbromide	0.10 - 5.0

5. AGRONOMIC AND ECOTOXICOLOGICAL ASPECTS OF HERBICIDE CONTAMINATION OF GROUNDWATER IN ITALY

M. VIGHI
Institute of Agricultural Entomology
University of Milan
Via Celoria 2
20133 Milano
ITALY

G. ZANIN
Institute of Agronomy
University of Padua
Via Gradenigo 6
35100 Padova
ITALY

1. The Detection of Herbicides in Groundwater in Northern Italy and the Ban on Atrazine

The Directive 80/778/EEC of 15 July 1980, on the quality of water for human consumption establishes a maximum admissible concentration of 0.1 μg/l for individual pesticides. This directive was incorporated into Italian law within DPCM No. 41 of 8 February 1985 and later within DPR No. 236 of 24 May 1988.

A survey of groundwater and public water supplies has been carried out which showed that herbicides, and in particular atrazine, were present in amounts above the admissible concentration established by the EEC, especially in Northern Italy.

Table 1 shows some of the results of the groundwater monitoring programme in three regions of Northern Italy (Lombardia, Piemonte and Veneto); the herbicide consumption of these three regions represents about 60% of the total amount used in Italy. In particular, due to the considerable maize culture there, the use of atrazine in these regions accounted for about 80% of the total national consumption up to 1988 [1].

Table 1 also contains a summary of the data collected from surveys carried out during 1989 by the different Regional Authorities. Due to the differences in the organisation of the surveys between the regions, the data obtained are not comparable.

From the Table it appears that the presence of atrazine above the EEC standard for drinking water occurs frequently, in about 20% of water supplies analysed in Lombardia and Piemonte and for more than 90% in Veneto. The extremely high figure for Veneto, based on a relatively small number of samples, is probably due to the selection of the sampling sites, which were concentrated in areas with a higher risk of contamination.

L. Bergman and D.M. Pugh (eds.), Environmental Toxicology, Economics and Institutions, 111-139.

Among the other herbicides detected, bentazone and simazine are frequently found, and various other compounds were present on a few occasions.

Table 1. Presence of herbicides in drinking water sources in the three regions of Northern Italy most liable to herbicide contamination. The number of water supplies or samples showing concentrations >0.1 ppb in the 1989 survey are given. Data are arranged differently for the various regions because of the variation in the type of information available.

	Lombardia Region*				Piemonte Region** Water samples at level		Veneto Region *** Water samples at level	
	PPW	Mun	Aqd	Mun	0.1-1 ppb	>1 ppb	0.1-1 ppb	>1 ppb
Alachlor	2	2	-	-	-	-	-	2
Ametryn	1	1	-	-	-	-	-	-
Atrazine	408	161	41	41	595	9	165	11
Bentazone	114	48	7	7	67	26	-	-
Metolachlor	2	2	-	-	-	-	-	1
Molinate	10	8	-	-	4	3	-	-
Propazine	6	6	-	-	-	-	-	-
Simazine	38	16	1	1	135	1	10	-
Terbumeton	-	-	-	-	7	-	-	-
Terbutylazine	6	4	-	-	13	-	-	2

*: Data refer to 775 municipalities, 2135 wells, 5340 analyses.
**: Data refer to 495 municipalities, 2536 wells, 2841 analyses.
***: Data refer to 46 municipalities, 175 wells, 14 surface waters, 189 analyses.
PPW: Public and Private Wells
Mun: Municipalities
Aqd: Aqueducts

Because of the detection of atrazine and other herbicides in drinking water supplies, several regulations were made, either containing derogations for these characteristics of water quality for human consumption, or providing for local or national restrictions on herbicide use.

A more detailed overview of the various regulations is given in Chapter 3 of this book. Briefly, after a series of local or temporary restrictions a national ban on the use of atrazine came into force in March 1990.

In terms of real risk for the exposed human population as a consequence of the consumption of water contaminated by atrazine, it is clear that the EEC limit for individual pesticides in drinking water is very conservative and, in many cases, orders of magnitude below a demonstrable toxicological threshold value. The philosophy of this part of the EEC Directive is based on the notion that products such as pesticides, (which are biocides by definition), should not be present in drinking water because this is a fundamental resource, intended for regular daily consumption.

The European Community policy has therefore followed "the general principle that drinking water should not contain any pesticides and that, given the limitations of current analytical techniques the limit of 0.1 μg/l for individual pesticides represents a practical zero value which can be used for enforcement purposes". In consequence, in EEC member states, control actions for the protection of drinking water are triggered not by a true toxicological limit but by a more stringent "legal" standard. This kind of standard establishes a very important principle in the protection of human health and compliance with it involves a long-term integrated strategy of management of pesticides and land use. On the other hand, this means that except for very few, strictly localised exceptions, the levels of atrazine found in groundwater in Northern Italy are unlikely to pose a health risk for the local populations as the toxicology-derived WHO Guideline for atrazine concentration in drinking water is up to 2 μg/l.

2. Agricultural Consequences of the Ban on Atrazine

2.1. POSSIBLE SUBSTITUTES FOR ATRAZINE

In Italy, before the legal restrictions came into force, atrazine was used in a variety of ways depending on the different weed communities present.

For 60-65% of the area in which maize was cultivated, atrazine was used mixed with monocotyledon herbicides (acetanilide, thiocarbamate herbicides, etc.); on 20-25% of the area it was used alone and on the remaining 5-10% it was not used because of the presence of resistant weed biotypes [2].

Table 2. Atrazine alternatives grouped according to previous atrazine use.

Atrazine + grass weed killers	Atrazine alone	No Atrazine
(1) Terbutylazine	Dicamba	(2) Dicamba
(1) Cyanazine	2,4D+MCPA	(2) 2,4D+MCPA
(1) Linuron	Bentazone	(1) Bentazone
	Pyridate	(2) Pyridate
(1) Pendimethalin	Bromoxynil	(2) Bromoxynil

(1) Pre-emergence
(2) Post-emergence

Since 1986-87, atrazine has been replaced with a number of other herbicides to control different assemblages of weeds, as shown in Table 2. The banning of atrazine has therefore "put back into play" almost all the selective herbicides for maize cultivation, and the selection of chemicals for weed control has become extremely specific in relation to particular pedoclimatic and floral conditions.

Post-emergence substitute herbicides have not produced negative agricultural effects; however, the situation for pre-emergence herbicides is very different as most have, under certain pedoclimatic conditions, a greatly reduced efficiency and a lesser persistence (Table 3). Thus their use implies both reductions in yield caused by their phytotoxicity and incomplete weed control as well as the need for repeated applications to eliminate late-emergence weeds or those which escaped the initial treatment.

The banning of atrazine, moreover, has made the control of *Echinochloa (= Panicum) crus-galli* (Cockspur panicum) a persistent weed in organic soils (around

100,000 hectares) more difficult; in these soils the control of this weed had been based on the use of a mixture of atrazine and oil at the post-emergence stage. The alternative would be to use thiocarbamate herbicides; however, soils repeatedly treated with these herbicides show an adaptation within a few years associated with an enhanced rate of herbicide degradation.

Table 3. Characteristics of atrazine and of alternative pre-emergence herbicides.

Herbicides	Persistence * (Months)	Efficiency	Selectivity
Atrazine	6-9	Very Good	Very Good
Terbutylazine	6-10	Good	Very Good
Cyanazine	2-3	Very Good	Good
Linuron	3-5	Average	Average
Pendimethalin	4-6	Average	Average

* The "agronomic" persistence is reported, i.e. the duration of residual activity against weeds or crops. It cannot be directly compared with the values for half life in soil (t1/2s) given in Table 7.

Furthermore, the substitution of atrazine by pre-emergence herbicides leads to a lesser flexibility of use; atrazine could be used pre-sowing, pre-emergence or post-emergence, during any phase of maize growth. Thus the substitutes involve a greater degree of farm organisation and require a much higher level of expertise in optimising weed treatment (Table 4).

On the other hand the banning of atrazine has had positive agronomic consequences, such as a reduction of the development of resistance phenomena and fewer problems with excessive persistence, thereby simplifying the planning of crop rotation.

Table 4. Agronomic disadvantages and weed control consequences from the substitution of atrazine by pre-emergence herbicides.

Disadvantages	Consequences	Results
Less selectivity	damage to maize	Yield reduction
Less reliability	lower weed control new weed species arrive	
Less persistence	development of late emergence weeds	Need for more treatments
Lack of control	e.g. barnyardgrass in organic soils	
Less user flexibility	more precision more complexity	Increase in professionalism

A new family of herbicides, with particularly favourable agronomic and ecotoxicological characteristics, will soon be widely available in Italy. These products are the sulfonylureas and their main characteristics are as follows

(1) they are enzyme inhibitors; the target enzyme is acetolactate synthetase (ALS) which is not present in mammals;

(2) they are active at very low doses (5-30 g/ha); 100-200 times lower than traditional herbicides;

(3) they have a very wide spectrum of activity;

(4) their persistence varies within the group; some have a very short half-life,

(5) their usage is very flexible (they are active at both pre- and post-emergence stages). Some examples are described in Table 5.

Table 5. Agronomic characteristics of some sulfonylureas.

Herbicides	Rate g/ha a.i.	Time of Application	Main Targets	Half-Life (Days)	
Thifensulfuron -methyl (DPX -M6316)	7.5	Post-emergence	Broadleaved Weeds	1 - 6	(1)
DPX - E9636	10 - 15	Post-emergence	Panicoid Grasses (monocotyledons)	0.3 - 4.6	(1)
CGA 136872	10 - 20	Post-emergence	Panicoid Grasses (monocotyledons)	17 - 30	(1)

(1) Dependent on pH and organic matter.

2.2. ALTERNATIVES TO CHEMICAL TREATMENT

The alternatives to chemical weed control are manual weeding, (almost always economically unjustifiable), and mechanical control. Selective flaming is restricted in its usage and it is only feasible with horticultural crops.

Mechanical weed control can only be carried out on crops sown in rows sufficiently widely spaced to allow the passage of the machine. Furthermore, the weeds close to the row must be eliminated in other ways, such as chemical control, or by hand, manual removal or hoeing.

"Finger weeders" have recently appeared on the market, which can operate specifically on high-density crops (e.g. wheat). These weeders, however, are insufficient for full control of the weeds, and also cause crop damage by the removal of "thinnings".

Weed problems can be controlled by agronomic techniques, combining those factors (sowing date, density and depth, spatial arrangement of the crop, choice of cultivar, suitable fertilisers) that favour the competitiveness of the crop. A well known adage is that "the best weed killer is always a good crop". The cost of good crop management should not be attributed to weed control. With agronomic techniques, it is also possible to restrict the occurrence of florisitic compensation (i.e. the predominance of a few species ecologically and physiologically similar to that of the crop). Crop rotation is the key to maintaining a balance in the weed infestation. However, there is not always compatibility between agronomic and economic requirements in the short term.

3. Ecotoxicological Consequences of the Ban of Atrazine

A hazard assessment and risk evaluation of a potentially dangerous chemical can be performed ideally by means of an integrated physico-chemical and ecotoxicological approach such as that schematically described in Figure 1. However, the information available for the herbicides of concern is not complete, in particular for the evaluation of a precise exposure concentration. This would require the development of "site specific" evaluative models and *ad hoc* research, which are not the aim of this present study.

Nevertheless it is possible to make some general evaluations which, even though they are rough approximations, will be sufficient for the assessment of the possible hazards

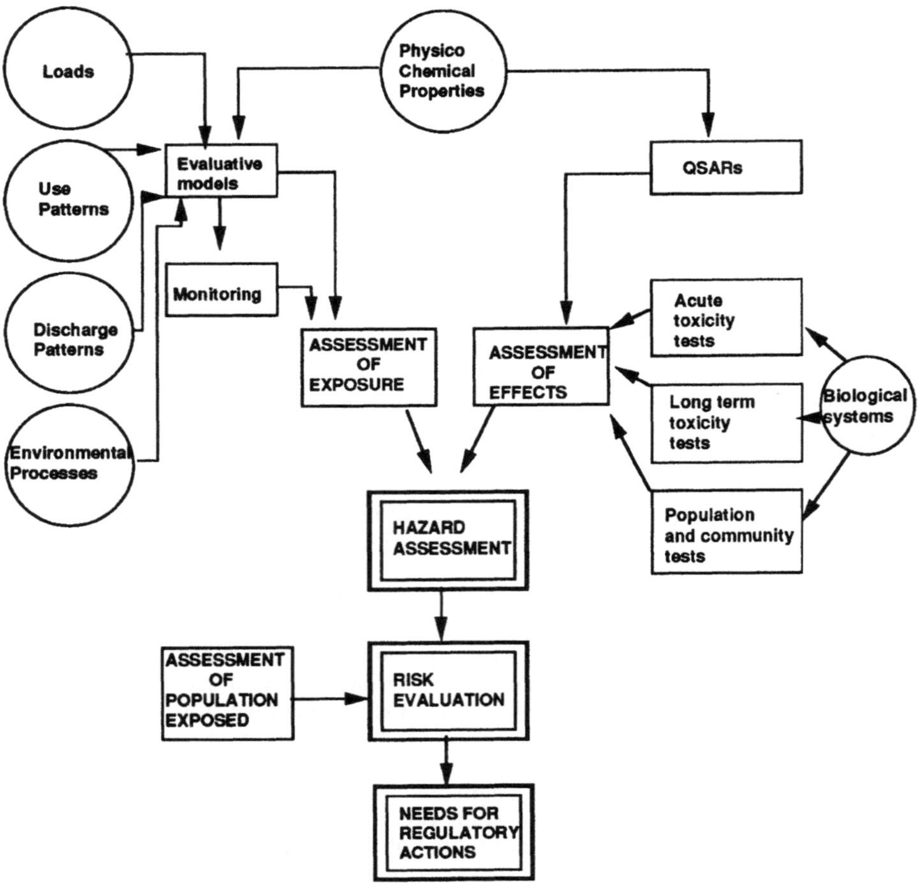

Figure 1. Integrated ecotoxicological approach for the study of potentially dangerous chemicals (modified from Vighi and Bacci, [3]).

for groundwater herbicide contamination, at least in comparative terms, either for the various molecular properties or for the situations that existed before the atrazine ban.

3.1. ENVIRONMENTAL DISTRIBUTION AND FATE OF ATRAZINE AND POSSIBLE SUBSTITUTES

As described in the previous section, a compete ban of atrazine in Italy would lead to an increase in the use of those herbicides listed in Table 2. While some of these were used heavily before the ban, others were not.(Table 6).

Table 6. Sales of herbicides in Italy during 2nd half 1986/1st half 1987 or in 1988 [4].

	1986/87	as tonnes/year	1988
Atrazine	301		634
Bentazone	95		122
Bromoxynil	0		7
Cyanazine	30		25
2,4 D	314		389
Dicamba	7		16
Linuron	271		284
MCPA	225		280
Pendimethalin	163		213
Pyridate	12		14
Terbutylazine	207		225
Alachlor	1109		1361
Metolachlor	544		712

Table 7. Main physico-chemical parameters and half life in soil (t1/2) for atrazine and its possible substitutes.

	M.W.	Water Solubility mg/l	Vapour Pressure mPa	log log Kow	Koc	Soil t1/2 days
Atrazine	215.7	30 [7]	0.04 [7[	0.04 [7]	2.4	80 (S)
Bentazone	240.3	500 [7]	0.46 [7]	-0.45 [7]	-0.66	48 [7]
Bromoxynil	276.9	130 [7]	0.6 [8]	2.6 (C)	2.4	15 [4]
Cyanazine	240.7	170 [7]	$2 \cdot 10^{-4}$ [7]	1.8 [9]	1.6	12-25 [7]
2,4D	221.0	620 [7]	1400 [10]	2.8 [11]	2.6	10 [7]
Dicamba	221.0	6500 [7]	4.5 [7]	2.15 (C)	1.9	14 [7]
Linuron	249.1	80 [7]	2 [7]	3.0 [7]	2.8	70 (S)
MCPA	200.6	800 [7]	0.2 [7]	2.5 (C)	2.3	20-50 [4]
Pendimethalin	281.3	0.3 [7]	4 [7]	5.2 [7]	5.0	90 (S)
Pyridate	378.9	15 [7]	$1.3 \cdot 10^{-4}$ [7]	4.9 (C)	4.7	10 [12]
Terbutylazine	229.7	8 [7]	0.15 [7]	3.0 [7]	2.8	70 (S)

Numbers in parentheses indicate references. C = calculated by means of the fragmental constant method [5]; S = selected (see Annex); Koc has been calculated from Kow [6].

The potential risk for groundwater from the use of these compounds, by comparison with that for atrazine, can be estimated by evaluating their environmental distribution and in particular their affinity for the water component, by means of predictive models.

The environmental distribution can be calculated by using the fugacity model [13, 14], modified by including the terrestrial plant biomass [15]. Table 7 shows the main physico-chemical parameters needed to operate the model and Table 8 shows the percent distribution between the main environmental compartments for the herbicides under study.

Table 8. Percent distribution of atrazine and possible substitutes between the main environmental compartments calculated by means of the fugacity model [14].

	Water	Air	Soil Sediment Susp. Sol.	Vegetal Biomass
Atrazine	91.1	$9 \cdot 10^{-3}$	8.4	0.47
Bentazone	99.9	$8 \cdot 10^{-3}$	0.01	0.04
Bromoxynil	91.2	0.04	8.3	0.36
Cyanazine	98.4	$9 \cdot 10^{-6}$	1.4	0.14
2,4 D	75.4	13.0	11.2	0.42
Dicamba	96.7	$5 \cdot 10^{-3}$	3.1	0.20
Linuron	80.7	0.7	18.5	0.62
MCPA	92.9	$2 \cdot 10^{-3}$	6.7	0.32
Pendimethalin	2.6	3.5	92.2	1.6
Pyridate	5.1	$5 \cdot 8^{-5}$	93.2	1.7
Terbutylazine	79.2	0.11	20.0	0.65

From Table 8 it appears that several products show a very high affinity for the water component. In particular, bentazone, bromoxynil, cyanazine, dicamba and MCPA have hydrophilic properties similar to, or greater than, that of atrazine; 2,4D, linuron and terbutylazine also show a high affinity for water, even though the values for percent distribution are slightly less. Only pendimethalin and pyridate appear to present a low risk for the water compartment.

It should be noted that there are some limitations to the use of the fugacity model for dissociated chemicals, as is the case for bromoxynil, 2,4D, dicamba and MCPA. Nevertheless, such dissociated anionic compounds are unlikely to be adsorbed in large amounts onto soil. Thus, in general, there will be a significant leaching of these compounds to groundwater.

On the other hand, in spite of their high aqueous solubility and low Koc, 2,4D and dicamba have been experimentally classified as "non-leachers" by the Californian Department of Food and Agriculture (CDFA) [16]. This is probably due more to the relatively short half-life in soil, than to an affinity for the soil component.

Another possible approach for evaluating the possibility that an organic chemical will reach groundwater lies in the so-called "leaching indices". Several of these indices have been put forward [17] with various levels of complexity and with different utility and limitations.

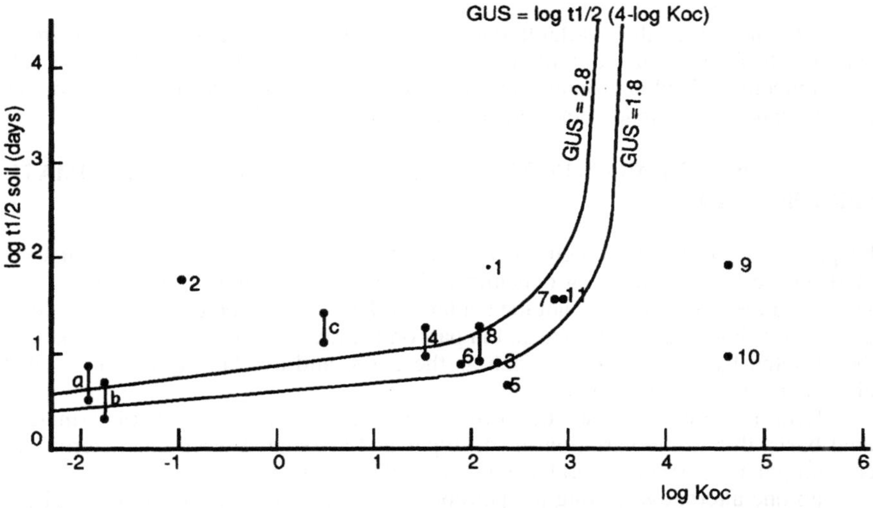

Figure 2. Use of the GUS index to predict the leaching potential for the herbicides studied. Solid lines indicate the boundaries of "leachers (GUS>2.8), "non-leachers" (GUS<1.8) and "transition" (1.8<GUS<2.8): Numbers refer to the herbicides listed in tables 6-8, letters refer to the sulfonylureas in Table 9.

In this study the GUS (Groundwater Ubiquity Score) index [12] will be used. This index, based on Koc and soil half-life (t1/2) is widely accepted and has the advantages of considerable simplicity and a limited input of data. The GUS index is calculated from the following equation.

$$GUS = \log t1/2 \, (4 - \log Koc)$$

There are two major limitations with this index. First the field application rates are not taken into account; thus, various products can be compared only on the basis of their potential capability for reaching groundwater, independently from the actual load. Secondly, reliable data on persistence are relatively difficult to obtain. A rough estimate of the t1/2 in soil, obtained from a survey of the literature, is given in Table 7.

Figure 2 shows the GUS indices for the herbicides considered in this study. This approach also has a limited application for dissociated compounds, because it takes into account only partition properties and does not include ionic binding to soil.

The index suggests that bentazone and cyanazine will be leachers, together with atrazine. Non-leachers include pendimethalin and pyridate, which is in agreement with the results of the fugacity model, and 2,4D, which is in agreement with the CDFA assessment. All the other compounds are classified in an "intermediate" group.

On the basis of using these two approaches it seems that of the possible substitutes for atrazine, only pendimethalin and pyridate and, with some reservations, 2,4D, can be considered as safe in the context of groundwater contamination. For all the other compounds there will be a risk of groundwater contamination to a greater or lesser extent. In particular, bentazone and the triazine compounds (especially cyanazine) appear to be products that give rise to some concern.

An analysis of the available data on groundwater contamination confirms this conclusion. In fact bentazone, with a national usage of about 1/5 that of atrazine (1988 data), is frequently found in groundwater, in particular in Lombardia, and other compounds which are relatively frequently detected are the triazine herbicides.

3.2. REACTION OF FARMERS TO THE ATRAZINE BAN AND ESTIMATE OF HERBICIDE LOADS

In the previous section the potential for leaching of the possible atrazine substitutes, selected on the basis of their agrochemical characteristics, has been compared. These evaluations did not take into account the loading of the different compounds which depends in part on their field application rate and in part on the reaction of farmers to the ban. This can be affected by several factors such as the costs, and availability on the market of substitutes, in agricultural practices, etc.

Even if there are substantial differences between the responses of farmers on a regional basis, due to the variation in soil types and associated weed characteristics and agricultural practices, some generalisations can still be made as follows:

- no one alternative among the possible substitutes for atrazine can be used alone; also, a combination of graminid (i.e. grasses) herbicides must always be applied;
- the most extensively used atrazine substitutes after the ban are terbutylazine (on about 80% of the maize hectarage), linuron (on about 10%) and pendimethalin (on about 10%);
- among the graminid herbicides used in combination with atrazine substitutes, the most common are alachlor and metolachlor, these compounds were used in large amounts before the ban, in combination with atrazine;
- other compounds may be used in small quantities, e.g. dicamba (post-emergence).

In summary, the most commonly used herbicide combinations after the ban of atrazine are the following:

terbutylazine + alachlor
terbutylazine + metolachlor — 80% of the maize hectarage

linuron + alachlor
linuron + metolachlor — 10% of the maize hectarage

pendimethalin + alachlor
pendimethalin + metolachlor
pendimethalin + linuron — 10% of the maize hectarage

From a consideration of the most likely reaction of farmers to the ban on atrazine, the following rough predictions of the herbicide load after the atrazine ban can be made. There will be:

A Large Increase in the Terbutylazine Load. Taking into account the application rate (about 1 kg A.I./ha) and the probable percentage of the maize area treated (around 70-80% of the total) the calculated increase will be around 300% of the 1986-1988 loading.

A Significant Increase in the Linuron Load. Assuming an application rate of the same order of terbutylazine and a maize area treated around 10-15% of the total, the calculated increase will be 50-75% of the 1986-1988 loading.

A Significant Increase in the Pendimethalin Load. Assuming an application rate of 1.8-2 kg A.I./ha and a treated area of 10% of the total, the calculated increase will be about 100% of the 1986-1988 loading.

A Moderate Increase in the Load of Graminid Herbicides, e.g. Alachlor and Metolachlor. Taking into account the application rates (2-2.4 and 1.2-1.5 kg A.I./ha for alachlor and metolachlor respectively) and assuming that after the ban the maize hectarage treated with these products will be increased by about 15-20%, the calculated increase will be 10-15% of the 1986-1988 loading for these compounds, which were used in large amounts on maize before the ban and on other crops.

A Slight Increase in the Load of Various Other Products. The outcome can be envisaged for other possible atrazine substitutes or for the graminid herbicides used in mixtures. However, even though in some particular cases this increase could be of local significance, on a wider basis it can be assumed to be negligible.

In conclusion the most significant consequence of the atrazine ban in terms of the use of substitutes appears at the time of writing to be an increase of the load of terbutylazine and linuron.

3.3. EVALUATION OF POSSIBLE HAZARD FROM HERBICIDES AFTER THE BAN OF ATRAZINE

Ecotoxicological profiles, based on critical reviews of the literature, have been constructed for atrazine and the most important substitutes; these are given in the Annex.

In the context of pesticide regulation and in particular the assessment of the possibility that a biocide will appear in groundwater at concentrations above the EEC water quality standard, only information on the load applied and its environmental distribution and fate are needed; this evaluation is independent of any toxicological characteristics. However, in order to provide reassurance that this standard is still more than adequate to protect human health, toxicological data have been included.

The information given in the Annex, together with a calculation of the possible loading and the available data from previous groundwater analyses, can be used to give the following outlines:

a. Terbutylazine. Shows an affinity for water greater than that of atrazine and a persistence in soil of the same order or slightly less. During the 1989 survey this compound was detected in groundwater at levels higher than the ECC limit for drinking water in all the three regions of Northern Italy involved in the exercise, although the frequency of occurrence was not very high.

Due to the large increase in the use of this compound, there will be a very high probability of a widespread occurrence of terbutylazine in groundwater at concentrations above the EEC limit. Thus terbutylazine is a compound of considerable concern, and it is likely to cause problems similar to those of atrazine, at least from an institutional point of view. From a toxicological point of view, there is no internationally recognised ADI or a recommended guideline for terbutylazine in drinking water, as there is for atrazine. Nevertheless, a comparison between terbutylazine and atrazine, based on the total available toxicity data for mammals and non-mammalian organisms, indicates a very similar potency for various organisms and endpoints. Thus it can reasonably be assumed, at least for a rough approximation, that terbutylazine is comparable to atrazine insofar as its toxic potential is concerned.

Therefore it can reasonably be expected that, as with atrazine, levels in groundwater dangerous to a population exposed to terbutylazine are not likely to occur.

b. Linuron. Shows an affinity for water slightly lower in comparison with atrazine and a persistence in soil of the same order or slightly less.

During the 1989 survey, linuron was not reported among the herbicides detected in groundwater, in spite of the relatively high loads used in 1986-1988. The predicted increase in the loading after the atrazine ban is significant but not very high. Thus linuron can be seen as a product of medium concern in relation to its occurrence in groundwater at levels above the EEC limit.

From a toxicological point view, a general comparison of the available information suggests that linuron is more or less similar to atrazine and terbutylazine, and for some organisms and endpoints it may be somewhat less toxic. Nevertheless, for linuron an increase in testicular tumours was observed in rats during two year feeding trials, albeit at very high doses (125 mg/kg diet). Therefore the possible occurrence of linuron in groundwater should be monitored carefully.

c. Pendimethalin. Shows a very low affinity for water and is classified as a non-leacher by the GUS index. Pendimethalin was not found in groundwater during previous surveys.

Taking into account the high WHO guideline value (17 μg/l), for drinking water, pendimethalin can definitely be classified as a compound of no concern with respect to groundwater, even though its loading may be doubled.

3.4. ECOTOXICOLOGICAL EVALUATION OF SULFONYLUREAS

The situation described so far relates to the reaction of the farmers to the atrazine ban and their use of alternative existing herbicides, and this can be assumed to be valid up to the 1991 growing season. However, as stated earlier, there is increasing interest in some relatively new sulfonylurea herbicides.

Since 1992 one such product has been available on the Italian market (thifensulfuron methyl) and another (DPX E9636), already available in some European countries, has been approved for use in Italy in 1993. At present, there are no reliable or official data available on the type and tonnage of herbicides used in maize in 1992 and 1993. Due to some problems, related to thifensulfuron methyl efficacy, it is reasonable to hypothesise that no significant changes occurred in 1992. Nevertheless, there is no doubt that the previous situation as described could be substantially changed and it is very probable that, in the near future, sulfonylureas will have a considerable and rapidly increasing share of the market.

Table 9 shows the main physico-chemical characteristics of the three sulfonylureas which are possible substitutes for atrazine, and their percentage distribution in the various environmental compartments as calculated by the fugacity model. All the compounds show a very high affinity for water.

Table 9. Physico-chemical parameters and environmental distribution of sulfonylureas. Numbers in parentheses indicate references. (C) = calculated.

	Physico-chemical parameters					% distribution			
	M.W.	Water sol g/l (pH7)	V.P. m Pa	log Kow (pH7)	log Koc (C)	Water	Air	Soil Sed Sus solids	Vegetal biomass
(DPX-M6316]	387.4 (33)	2.4 (33)	$17.3.10^{-6}$ (33)	-1.6 (33)	-1.8	99.96	9.10^{-8}	6.10^{-4}	3.10^{-2}
DPX-E9636	43.4 (3)	7.3 (3)	1.10^{-6} (C)	-1.5 (3)	-1.7	99.96	2.10^{-6}	7.10^{-4}	3.10^{-2}
CGA-136872	468.3 (1)	0.07 (1)	1.10^{-6}	1(C)	0.8	99.70	$2.3.10^{-4}$	0.22	$7.3.10^{-2}$

DPX-M6316 = Thifensulfuron-methyl

Table 10. Toxicological data available for sulfonylureas (7, 18, 19, 20).

Mammalian Toxicity

	Oral LD50 mg/kg	Percutaneous LD50 mg/kg	Inhalation LC50 mg/l	NOEL mg/kg diet	Teratogenesis
Thifensulfuron-methyl (DPX-M6316)	>5000	>2000	>7.9	25	NOEL 200 mg/kg diet
DPX E9636	>5000	>2000	-	-	-
CGA 136'872	>5000	>2000	>4.8	-	-

Non Mammalian Toxicity

	Terrestrial acute LD50			Aquatic acute LC50		
	Birds mg/kg	Bees ug/bee	Earthworms mg/kg	Fish mg/l	Daphnia mg/l	Ames test
Thifensulfuron -methyl (DPX-M6316)	>2510	>12.5	>2000	>100	>1000	neg
DPX E9636	-	-	-	-	-	neg
CGA 136'872	*	-	-	*	*	neg

- : no data available

* : indicated as "practically non toxic" but no quantitative data are reported.

On the basis of the GUS index thifensulfuron and DPX-E9636 can be classified as "transition" compounds, whereas CGA-136872, because of its relatively high persistence, is classified as "leacher" (Figure 2).

On the other hand, if we take into account the extremely low application rates for sulfonylureas which are about 100 times lower than traditional herbicides (Table 5), the possibility that they will occur in groundwater at concentrations above the EEC limit can be assumed to be negligible, in particular for the first two compounds. Also, despite their very high potency on target plants, their toxicity to animals appears to be relatively low.

An ecotoxicological profile for the compound at present used in Italy (thifensulfuron methyl) is given in the Annex. A comparison with atrazine and its most important substitutes indicate that, for various organisms and endpoints, the toxicity of this sulfonylurea is broadly comparable (at least a similar order of magnitude) and, in some cases, orders of magnitude lower (e.g. bees, aquatic animals).

The few data available for the other two compounds indicate that they have a low toxicity, at least for mammals (Table 10) and a negative response to the Ames test [18, 19].

In conclusion, the new sulfonylureas seem to be very promising alternatives to atrazine, not only from the agronomical point of view, but also because of their ecotoxicological properties. In developing a strategy for the protection of groundwater from herbicide contamination, the consequences of the use of these new products should be evaluated.

Because of the relatively limited experience with these compounds, the reliability of the available information should be critically assessed, in particular the data on persistence. Also, it is known that the variability of some physico chemical properties of these compounds in relation to changes in environmental parameters is particularly high. For example, water solubility and Kow can differ by two orders of magnitude between pH5 and 7, which may have a significant effect on the environmental distribution and fate of these herbicides.

Therefore, although the existing data on sulfonylureas are encouraging their ecotoxicological properties should be carefully checked.

4. Hydrogeological and Pedological Aspects

The potential for a chemical substance to reach groundwater, and the transport paths involved, depends not only on physico-chemical properties of the molecule but, also of course, on the hydrogeological features of the area and some characteristics of the soil (e.g. organic carbon content, soil density, functional porosity, etc.).

Therefore, for a complete risk assessment of the possibility of herbicides reaching groundwater and in order to construct a proper strategy for control measures in a certain region, information on the hydrogeological and pedological features is required.

In particular, agricultural areas within the region can be separated in terms of the vulnerability of their aquifers as follows:

(a) areas where herbicides could be used without particular restrictions;

(b) areas where "leaching" herbicides could be used only in exceptional cases, and with a restriction on the load applied;

(c) areas where only "non-leaching" herbicides can be used and with some control of the load applied;

(d) areas where the use of herbicides (and perhaps all pesticides) should be completely forbidden because of the vulnerability of the aquifers and their strategic value as drinking water supplies.

The widespread groundwater contamination in Northern Italy, and in particular in the Po Valley, has been caused by the infiltration of water through very permeable soils and strata, or through poorly cultivated soils or those with a low organic carbon content.

The information needed for such a classification of agricultural areas (and the subsequent imposition of controls) is as follows:

- hydraulic balance of soils;
- permeability of surface soils;
- organic content of soils;
- water-dispersive characteristics of the soil particles;
- hydrogeological structure (e.g. depth of aquifers, pathways and thicknesses of strata with different permeabilities, (and the piezometry of aquifers);
- influence of cationic exchange capability of different clay materials and various humus types on the equilibrium reached with dissociated and undissociated contaminants;
- distribution of the crop types;
- herbicides used and their characteristics (physico-chemical properties, degradability, etc.).

Because this information is not yet complete for the Po Valley and because various characteristics show a different distribution within the region, specific maps of the area should be prepared by *ad hoc* studies to provide a general picture of the hydrogeological condition there.

These maps (drawn to a scale of not less than 1/500000) will enable areas to be defined where specific herbicide controls should be implemented.

ANNEX

Ecotoxicological Profiles of those Herbicides of Major Interest

This Annex contains condensed ecotoxicological profiles constructed for atrazine, linuron, pendimethalin, terbutylazine and thifensulfuron methyl. The information is presented in the following sequence and is the result of a critical review of the available literature.

Substance Identification. Chemical Abstracts name, CAS Registry number, structure and molecular formulae are reported.

Physico-Chemical Properties. Major properties of environmental relevance are reported.

Agronomic Aspects. The main uses and application rates are reported.

Toxicology. Toxicity data for various endpoints for mammals and for non-mammalian terrestrial and aquatic animals are reported.

If available an Admissible Daily Intake (ADI) and a guideline for drinking water, as well as information on mutagenesis, oncogenesis and teratogenesis are given.

Experimental values for the bioconcentration factor (BCF) have been taken from the literature; also a potential BCF has been calculated from the Kow [21].

Environmental Fate. A digest of information on persistence in soil, water and air compartments is given.

Environmental Distribution. The Level 1 fugacity model [14], modified by the inclusion of the terrestrial plant compartment [22], has been applied.

GUS Index. The Groundwater Ubiquity Score (GUS) index [12] has been taken from the literature or calculated on the basis of Koc (calculated from Kow according [6]) and the soil half-life.

1. Atrazine

1.1. SUBSTANCE IDENTIFICATION

C.A. 6-chloro-N-ethyl-N'-(1-methyl ethyl)-1,3,5-triazine-2,4-diamine.

Structure.

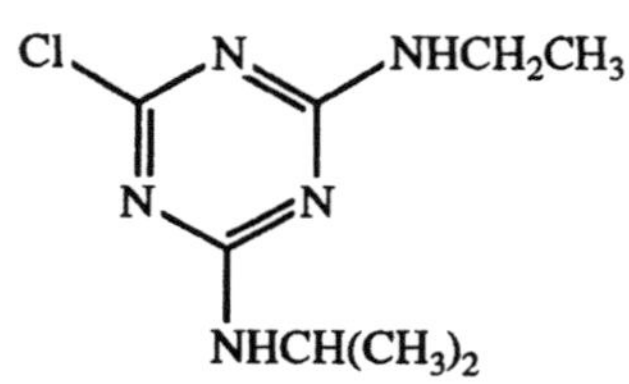

CAS Registry Number. 1912-24-9

Molecular Formula. $C_8H_{14}ClN_5$

1.2. PHYSICO-CHEMICAL PROPERTIES

		Ref.
Melting Point °C:	175-177	[7]
Molecular Weight:	215.7	[7]
Log Octanol/Water Partition Coefficient:	2.61	[9]
Water Solubility mg/l:	30	[7]
Vapour Pressure Pa:	4.0E-05	[7]
Volatilization coefficient (Hartley's law):	0.00059	

1.3. USE AND APPLICATION RATES

Atrazine is a 1,3,5-triazine herbicide typically used in maize crop against *Amaranthus, Capsella, Chaenopodium, Echinocloa, Geranium, Matricaria, Mercurialis, Polygonum, Plantago, Poa, Portulaca, Solanum, Stellaria, Urtica, Veronica.* It is practically innocuous to perennial plants, like, *Cynodon, Sorghum halepense, Convolvolus arevensis*, because of their deep root systems (rhizomes). Some crops are sensitive to atrazine, therefore sowing should be delayed for at least five months (wheat) or 10 months (sugar beet, tobacco) [23].

Application rates (data calculated from the rate given for 47.5% A.I. formulations):

2.0 kg A.I./ha in 6-10 hl water, before sowing.

2.0 kg A.I./ha in 4-6 hl, when maize is 20-40 cm high [23].

1.4. TOXICOLOGY

1.4.1. Mammalian Toxicity

Acute. oral LD_{50} rats 1869-3080 mg (T.C.)/kg; inhalation LD_{50} rats (1-hr) >0.71 mg/l [7]; percutaneous LD_{50} rabbits 9300 mg (80% w.p.)/kg [24].

NOEL. (2-year oral) rats 8 mg/kg d.; dogs 5 mg/kg d. [7].

1.4.2. Non-Mammalian Toxicity:

Terrestrial. Invertebrates: acute oral LD_{50} bees >0.16 mg/bee [25]. Birds: acute oral LD_{50} quail 5760 mg/kg diet, ducks 19650 mg/kg diet [7].

Aquatic. Fish: acute LC_{50} (96-hrs) bluegill 16 mg/l; carp 76->100 mg/l [7]. Crustaceans: acute LC_{50} (96-hrs) 5.7-6.9 mg/l [25].

1.4.3. BCF (fish)

Experimental: *Cottus bairdi* 2.0; *Leuciscus idus* 1.0; *Ictalurus melas* 0.5 [11].

Calculated from Kow: [7].

Other organisms: (experimental) phytoplankton 10-83; reptiles 2-15 [26].

1.4.4. ADI.
0.7 µg/kg [27]

1.4.5. Drinking Water Guideline
2 µg/l [27]

1.4.6. Oncogenesis, Carcinogenesis, Teratogenesis
The limited information available suggests that atrazine might be a weak, non-genotoxic carcinogen in rat. A NOEL of 0.7 mg/kg could be assumed [27].

1.5. ENVIRONMENTAL FATE

1.5.1. Terrestrial
Atrazine has a half-life ranging from 48 d [22], to 64 d [28], to 74 d [12]. Other authors report a t1/2 of 60-160 d [7], 224 d (pH=4; 25°C, without additives) and 1.73 d (2% of organic matter) [11]. Particularly rich organic soils are unsuitable for treatment with atrazine, which is inactivated by binding to fulvic acids. Generic persistence is estimated to be between 4-12 months [23].

1.5.2. Aquatic
Half-life values in water range from 30 d [23] to 77-86 d [29]. In sediments, the half-life is about 75 d [29]. Other authors report t1/2 values from 34.8 d (pH=2.9), to 174 d (pH=4.5), 398 d (pH=6.0), and 742 d (pH-7.0) [11].

1.5.3. Atmospheric
Only a single value for t1/2 available, i.e. 2.6 hrs at 25°C [11].

1.6. ENVIRONMENTAL DISTRIBUTION

Use of the Mackay model suggest the following percentage distribution between compartments:

Compartment	%
Air	9.063E-03
Water	91.084E+00
Soil	4.412E+00
Sediments	4.118E+00
Suspended Solids	6.864E-03
Aquatic Biomass	3.006E-03
Root Biomass	0.150E+00
Stem Biomass	9.931E-02
Foliage Biomass	0.117E+00
Vegetal Biomass Total	0.366E+00

1.7. GUS INDEX

Atrazine can be considered as a probable leacher, with a GUS Index of 3.06 calculated on the basis of a selected t1/2 value of 80 days.

2. Linuron

2.1. SUBSTANCE IDENTIFICATION

C.A. N'-(3,4-dichlorphenyl)-N-methoxy-N-methyl urea

Structure.

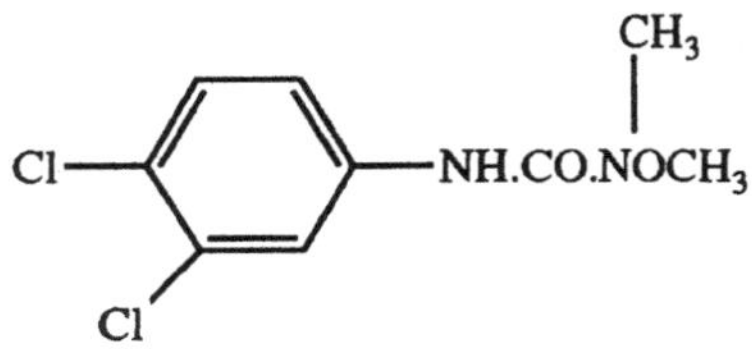

CAS Registry Number. 330-55-2

Molecular Formula $C_9H_{10}Cl_2N_2O_2$

2.2. PHYSICO-CHEMICAL PROPERTIES

		Ref.
Melting Point °C:	93-95	[7]
Molecular Weight:	249.1	[7]
Log Octanol/Water Partition Coefficient:	3.0	[7]
Water Solubility mg/l:	81	[7]
Vapour Pressure Pa:	2.0E-03	[7]
Volatilization coefficient (Hartley's law):	0.032	

2.3. USES AND APPLICATION RATES

Linuron is a urea herbicide, and a photosynthetic electron transport inhibitor. It acts principally by root absorption. It can be used on maize, beans, peas, cotton, soyabean, potatoes, winter wheat against *Amaranthus, r., Capsella b.p., Chenopodium spp., Digitaria s., Matricaria c., Mercurialis a., Portulaca o., Setaria spp., Sinapis a., Solanum n.* [23].

Application rates (data are calculated from the rates given for 47.5% A.I. formulations):

Maize, strawberry, carrot, potato: about 0.7-1.2 kg A.I./ha in 4-6 hl water.

Fennel: 0.25-0.5 kg A.I./ha in 4-6 hl water.

Rice field embankments: 2 kg A.I./ha in 4-6 hl water [23].

2.4. TOXICOLOGY

2.4.1. Mammalian Toxicity

Acute. Oral LD_{50} rats 1254 mg/kg [24]; dogs 500 mg/kg [7]. Percutaneous LD_{50} rabbits >5000 mg/kg [7].

NOEL. (2-y.f.t.) rats and dogs 125 mg/kg diet [7].

2.4.2. Non-Mammalian Toxicity

Terrestrial. Invertebrates: acute oral LD_{50} bees >0.16 mg/bee (oral), >0.016 mg/bee (contact) [25]. Birds: acute oral LD_{50} duck and quail >5000 mg/kg diet.

Aquatic. Fish: acute LC_{50} (96-hrs) bluegill and rainbow trout 16 mg/l water [7], other data for fish 3.2-6.5 mg/l [25]. Crustaceans, 0.75 mg/l [25].

2.4.3. BCF (fish)

Experimental: no data. Calculated from Kow: 71.

2.4.4. ADI

No data available.

2.4.5. Oncogenesis, Carcinogenesis, Teratogenesis

At high doses (>125 mg/kg diet) an increase in testicular tumours has been observed in rats. A NOEL of 50 mg/kg diet could be assumed [24].

2.5. ENVIRONMENTAL FATE

2.5.1. Terrestrial

Several values for half-life have been reported, ranging from 38-67 d [7] to 60-150 d [24]. Other authors give t1/2 values of 131 [25] and 139 d [17]. Therefore, values between 28 and 178 d [30] may be considered reasonable. In anaerobic conditions, the t1/2 can extend to 1.95 year [30].

2.5.2. Aquatic

Similarly, there are wide variations in the half-life values for linuron in water, which range from 24 to 239 d [25, 17]. Some authors make a distinction between surface water (t1/2 = 28-179 d) and ground water (t1/2 = 56-356 d) [30].

2.5.3. Atmospheric

Linuron has a short half-life, reported values ranging from 0.49 to 4.9 hours [30].

2.6. ENVIRONMENTAL DISTRIBUTION

Use of the Mackay model suggest the following percentage distribution between compartments:

Compartment	%
Air	0.172E+00
Water	80.647E+00
Soil	9.590E+00
Sediments	8.951E+00
Suspended Solids	1.492E-02
Aquatic Biomass	5.709E-03
Root Biomass	0.237E+00
Stem Biomass	0.129E+00
Foliage Biomass	0.254E+00
Vegetal Biomass Total	0.620E+00

2.7. GUS INDEX

A value of 3.0 (i.e. probable leacher) is obtained [17]. Using the different values given for the half-life for linuron, it is possible to define a range of GUS Index from 1.9 (t1/2=38 d [7] to 2.7 (t1/2=178 d [30]). Within these extremes linuron can be considered as a transitional leacher.

3. Pendimethalin

3.1. SUBSTANCE IDENTIFICATION

C.A. N-(1-ethylpropyl)-3,4-dimethyl-2,6-dinitrobenzenamine

Structure.

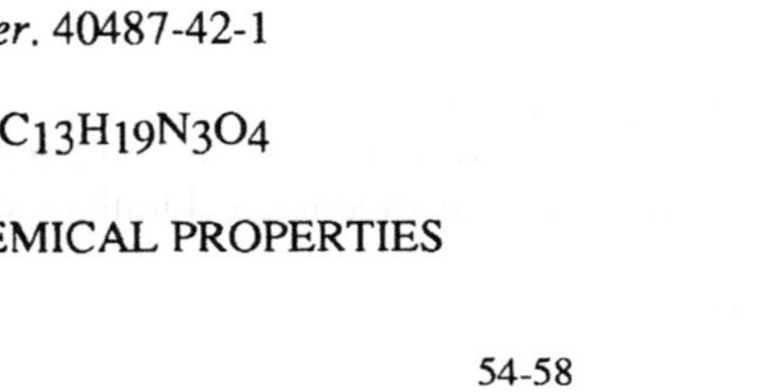

CAS Registry Number. 40487-42-1

Molecular Formula $C_{13}H_{19}N_3O_4$

3.2. PHYSICO-CHEMICAL PROPERTIES

		Ref.
Melting Point °C:	54-58	[7]
Molecular Weight:	281.3	[7]
Log Octanol/Water Partition Coefficient:	5.18	[7]
Water Solubility mg/l:	0.3	[7]
Vapour Pressure Pa:	4.0E-03	[7]
Volatilization coefficient (Hartley's law):	0.067	

3.3. USE AND APPLICATION RATES

Pendimethalin is a 2,6-dinitroaniline cell-division inhibitor herbicide. It can be used on maize, potato, pea, tobacco, *Liliaceae*, wheat and barley against different annual graminid and dicotyledonous weeds such as *Alopecurus m.*, *Fumaria o.*, *Poa a.*, *Polygonum a.*, *Ranunculus* spp., *Raphanus r.*, *Matricaria c.* [23].

Application rates (data are calculated from the rate given for 31.7% A.I. formulations): Carrot and soyabean: 1.2-1.8 kg/ha. Wheat, barley and maize crop: 1.8-2 kg/ha [23].

3.4. TOXICOLOGY

3.4.1. Mammalian Toxicity

Acute. oral LD_{50} rats 900-1125 mg A.I./kg; dogs >4500 mg/kg [7]. Percutaneous LD_{50} rabbits >4500 mg/kg [7].

NOEL. (2-y.f.t.) rats 100 mg/kg diet [7].

3.4.2. Non-Mammalian Toxicity

Terrestrial. Invertebrates: acute oral LD_{50} bees > 0.045 mg/bee (contact) [7]. Birds: acute oral LD_{50} quail 3770 mg/kg [7].

Aquatic. Fish: acute LC_{50} (96-hrs) Bluegill 0.14-0.19 mg/l water [25]. Crustaceans: acute LC_{50} (96-hrs) 0.08 mg/l [25]. Others: no data.

3.4.3. BCF (fish)

Experimental: no data. Calculated from Kow: 5012.

3.4.4. ADI

5 µg/kg [27].

3.4.5. Drinking Water Guideline

17 µg/l [27].

3.4.6. Oncogenesis, Carcinogenesis, Teratogenesis

Long term studies in mice and rats do not provide evidence of pendimethalin carcinogenicity; however these studies have methodological limitations [27].

3.5. ENVIRONMENTAL FATE

3.5.1. Terrestrial

Half-life value range from 30-90 d [7] to 171 d [25, 17] depending on the soil characteristics.

3.5.2. Aquatic

In water the half-life is <21 d [7]. Other authors report a t1/2 for hydrolysis of >28 d [25].

3.5.3. Atmospheric
Pendimethalin has a half-life (photolysis) of <7 d [25].

3.6. ENVIRONMENTAL DISTRIBUTION

Use of the Mackay model suggests the following percentage distribution between compartments:

Compartment	%
Air	3.439E+00
Water	2.640E+00
Soil	47.701E+00
Sediments	44.521E+00
Suspended Solids	7.420E-02
Aquatic Biomass	1.337E-02
Root Biomass	0.329E+00
Stem Biomass	7.117E-03
Foliage Biomass	1.265E+00
Vegetal Biomass Total	1.601E+00

3.7. GUS INDEX

With calculated GUS indices -1.43 and 1.83, pendimethalin can be considered to be a non-leacher.

4. Terbutylazine

4.1. SUBSTANCE IDENTIFACTION

C.A. 6-chloro-N-(1,1-dimethylethyl)-N'-ethyl-1,3,5-triazine-2,4-diamine.

Structure.

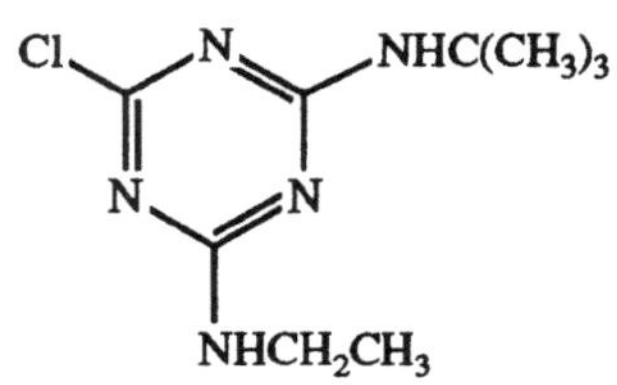

CAS Registry Number. 5915-41-3

Molecular Formula. $C_9H_{16}ClN_5$

4.2. PHYSICO-CHEMICAL PROPERTIES

		Ref.
Melting Point °C:	177-179	[7]
Molecular Weight:	229.7	[7]
Log Octanol/Water Partition Coefficient:	3.04	[7]
Water Solubility mg/l:	8.5	[7]
Vapour Pressure Pa:	1.5E-04	[7]
Volatilization coefficient (Hartley's law):	0.0023	

4.3. USE AND APPLICATION RATES

Terbutylazine is a 1,3,5-triazine herbicide, inhibiting photosynthetic electron transport and is absorbed by the root system. It can be used in orchards, viticulture, citrus groves, and urban areas, to control annual graminid and dicotyledonous weeds [23].

Application Rates: Fructi-viticulture and citrus: 1-1.5; 2.5-4 kg A.I./ha on young and fully grown plants respectively. Cereals: 0.8-1.2 kg A.I./ha Urban areas: 5-10 kg A.I./ha [23].

4.4. TOXICOLOGY

4.4.1. Mammalian Toxicity

Acute. oral LD_{50} rats 2000 mg (TC)/kg; percutaneous LD_{50} rats >3000 mg/kg; inhalation LC_{50} rats 3.51 [7].

NOEL. (90 day oral) rats 3.5 mg/kg diet [7].

4.4.2. Non-Mammalian Toxicity

Terrestrial. No data.

Aquatic. Fish: acute LC_{50} (96-hrs) rainbow trout 4.6 mg/l, carp 66 mg/l; bluegill 52 mg/l [7]. Crustaceans: acute LC_{50} *Daphnia m.* 5.5 mg/l (calculated from QSAR) [31].

4.4.3. BCF (fish)

Experimental: No data.

Calculated from Kow: 76.

4.4.4. ADI

No available data.

4.5. ENVIRONMENTAL FATE

4.5.1. Terrestrial

Various values for half-life are reported, ranging from 30-90 d [7] to 114 d [17], with a general persistence of 4-12 months [23].

4.5.2. Aquatic
The half-life for hydrolysis (calculated) is lowest at very high or low pH values, e.g. 8 d (pH-1.0), 86 d (pH=5.0), >200 d (pH=9.0), 12 d (pH=13.0) [7].

4.5.3. Atmospheric
No data.

4.6. ENVIRONMENTAL DISTRIBUTION

Use of the Mackay model suggest the following percentage distribution between compartments:

Compartment	%
Air	0.111E+00
Water	79.239+00
Soil	10.332E+00
Sediments	9.643E+00
Suspended Solids	1.607E-02
Aquatic Biomass	6.066E+03
Root Biomass	0.248E+00
Stem Biomass	0.131E+00
Foliage Biomass	0.274E+00
Vegetal Biomass Total	0.653E+00

4.7. GUS INDEX

With a GUS Index of 3.11, terbutylazine can be considered to be a probable leacher [17]. A calculated value of 2.2, based on a selected T1/2 of 70 days, would classify the chemical as a transitional leacher.

5. Thifensulfuron-Methyl

5.1. SUBSTANCE IDENTIFICATION

C.A. 3-[(4-methoxy-6-methyl-1,3,5-triazine-2-yl)amino-carbonyl-amino-sulfonyl]-2-thiophene carboxylic acid

Structure.

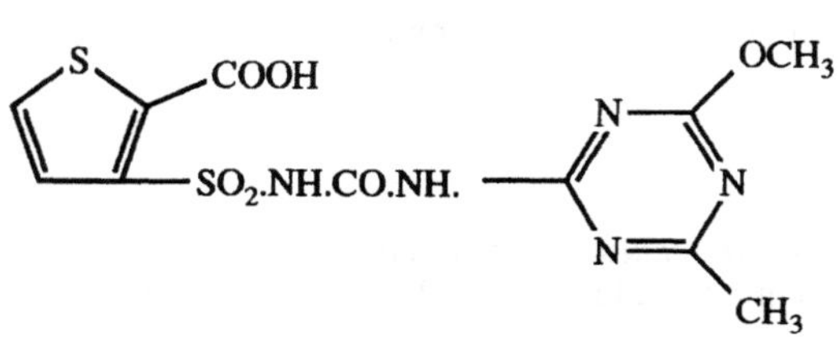

CAS Registry Number. 79277-27-3

Molecular Formula. $C_{12}H_{13}N_5O_6S_2$

5.2. PHYSICO-CHEMICAL

		Ref.
Melting Point °C:	186	[7]
Molecular Weight:	387.4	[7]
Log Octanol/Water Partition Coefficient:	0.5 (pH=5) -1.60 (pH=7)	[7]
Water Solubility mg/l:	24 (pH=4) 2400 (pH=6)	[7]
Vapour Pressure Pa:	1.7E-08	[7]
Volatilization coefficient (Hartley's law):	2.84E-07	

5.3. USE AND APPLICATION RATES

Thifensulfuron-methyl is a sulfonylurea herbicide, branched chain amino acid synthesis inhibitor. It is used to control broad-leaved weeds in cereals [7].

Application rates: for post-emergence: 20-40 g A.I./ha [7].

5.4. TOXICOLOGY

5.4.1. Mammalian Toxicity

Acute. oral LD_{50} rats 5000 mg/kg; percutaneous LD_{50} rats 2000 mg/kg [7]; inhalation LC_{50} rats 7.9 mg/l.

NOEL. (2-year oral) rats 25 mg/kg diet [7].

5.4.2. Non-Mammalian Toxicity

Terrestrial. Invertebrates: acute oral LD_{50} bees >12.5 μg/bee; LD_{50} earthworm >2000 mg/kg. Birds: acute oral LD_{50} ducks >2510 mg/kg [7].

Aquatic. Fish: acute LC_{50} (96-hrs) rainbow trout and bluegill >100 mg/l [7]. Crustaceans: acute LC_{50} (48-hrs) *Daphnia* 1000 mg/l [7]..

5.4.3. BCF (fish)

Experimental: no data. Calculated from Kow: 0.008.

5.4.4. ADI.

No available data.

5.4.5. Oncogenesis, Carcinogenesis, Teratogenesis

No teratogenic effects in rats and rabbits to 200 mg/kg diet [7].

5.5. ENVIRONMENTAL FATE

5.5.1. Terrestrial
Half-life values for thifensulfuron-methyl in soils range between 1 and 4 d [7].

5.5.2. Aquatic
Half-life in water (hydrolysis) varies with pH ranging from 4.7 hrs (pH=3), 38 hrs (pH=5), 250 hrs (pH=7), to 11 hrs (H=9) [7].

5.5.3. Atmospheric
No available data.

5.6. ENVIRONMENTAL DISTRIBUTION

Use of the Mackay model suggest the following percentage distribution between compartments:

Compartment	%
Air	9.490E-08
Water	99.965E+00
Soil	2.986E-04
Sediments	2.787E-04
Suspended Solids	4.645E-07
Aquatic Biomass	8.707E-07
Root Biomass	3.455E-02
Stem Biomass	3.129E-04
Foliage Biomass	7.920E-06
Vegetal Biomass Total	3.487E-02

5.7. GUS INDEX

With a GUS Index of 3.47, thifensulfuron-methyl can be considered as a probable leacher [17]. This classification is confirmed by a calculated value of 3.37.

REFERENCES

1. Ministero della Sanita' (1990): *Vendite di sostanze diserbanti. Anno 1988*. Ministero della Sanita. D.G. Igiene Alimenti e Nutrizione. Roma.
2. Zanin, G., Molle, I., Vazzana, C. (1984): Distribution et extensione de dicotyledones adventices a l'atrazine en Italie. *7eme Coll. Int. Ecologie, Biologie, Systematique mauvaises herbes*. EWRS, 273-280.
3. Vighi, M., Bacci, E. (1988): The Mediterranean sea: environmental impact of chemicals. *Proc 1st European Conference on Ecotoxicology, Copenhagen*, 333-344.
4. Funari, E., Bastone, A., Bottoni, P., Carbone, S., Donati, L., Marcello, I., Camoni, I., Giuliano, G. (1991): Il rischio delle contaminazione da erbicidi dell risorse idropotabili. La situazione in Italia. *Rapporti ISTISAN 91/2, ISSN-0391-1675*. Roma.

5. Hansch, C., Leo, A.J. (1979): *Substituent constants for correlation analysis in chemistry and biology*, John Wiley, New York.
6. Karickoff, S.W., Brown, D.S., Scott, T.A. (1979): Sorption of hydrophobic pollutants on natural sediment, *Water Research* 13, 241-248.
7. Worthing, C.R., Hance, R.J. (Eds.) (1991): *The pesticide manual, 9th edition.* The British Crop Protection Council, Farnham.
8. Weed Science Society of America (1989); *Herbicide Handbook*, W.S.S.A., Champaign, Illinois.
9. Finizio, A., Di Guardo, A., Arnoldi, A., Vighi, M. (1991): Different approaches for the evaluation of Kow for S-triazine herbicides, *Chemosphere* 23;801-811.
10. Suntio, L.R., Shiu, W.Y., Mackay, D., Sieber, J.N., Glotfelty, D. (1988). Critical review of Henry's law constants for pesticides. *Rev. Environ. Contam. Toxicol.* 103, 1-59.
11. Howard, P. (1991): *Fate and exposure data for organic chemicals. Vol. III, Pesticides*, Lewis Publishers Inc. Chelsea.
12. Gustafson, D.I. (1989): Groundwater ubiqity score: a simple method for assessing pesticide leachability. *Environ. Toxicol. Chem.* 8, 339-357.
13. Mackay, D. (1979): Finding fugacity feasible, *Environ. Sci. Technol.* 13, 1218-1223.
14. Mackay, D., Paterson, S. (1981): Calculating fugacity, *Environ. Sci. Technol.* 15, 1006-1014.
15. Bacci, E., Calamari, D., Gaggi, C., Vighi, M. (1990): Bioconcentration of organic chemical vapours in plant leaves: experimental measurements and correlation, *Environ. Sci. Technol.*. 24, 885-889.
16. Wilkerson, M.R., Kim, K.D. (1986): *The pesticide contamination prevention act: setting specific numerical values.* California Dept. of Food and Agriculture, Environmental Monitoring and Pest Management, Sacramento, CA.
17. Fielding, M., Barcelo, D., Helweg, A., Galassi, S., Torstensson, L., Van Zoonen, P., Wolter, R., Angeletti, G. (1991): Pesticides in ground and drinking water. *Water Pollution Research Reports 27* (Commission of the European Communities, Directorate - General for Science, Research and Development), Brussels.
18. Bassi, A., Bencivelli, A., Fabiani, G.P., Gamberini, C., Massasso, W., Salomone, M.C., Turchiarelli, V. (1990): DPX-E9636. Nuovo graminicida di postemergenza del mais ad ampio spettro e ridotta persistenza, *Atti Giornate Fitopatologiche*, 1, 125-132.
19. Airoldi, M., Casola, F., Filippi, G., Ruberti, R., Saporiti, G. (1992): Un nuovo erbicida di post-emergenza per il diserbo del mais, *Atti Giornate Fitopatologiche*. 1, 85-92.
20. Bassi, A., Bencivelli, A., Fabiani, G.P., Gamberini, C., Mossasso, W., Salomone, M.C., Turchiarelli, V. (1990): Thifensulfuron-methyl (Harmony) e le sue associazioni nel diserbo di postemergenza del mais. *Atti Giornate Fitopatologiche*, 1, 113-124.
21. Veith, G.D.., De Foe, D.L., Bergstedt, B.V. (1979): Measuring and estimating the bioconcentration factor of chemicals in fish, *J. Fish Res. Board Can.* 36, 1040-1048.
22. Bacci, E., Renzoni, A., Gaggi, C., Calamari, D., Franchi, A., Vighi, M. (1989): Models, field studies, laboratory experiments: an integrated approach to evaluate the environmental fate of atrazine (s-triazine herbicide), *Agriculture, Ecosystem and Environment*, 27, 513-522.
23. Muccinelli, M. (1987): *Prontuario dei Fitofarmaci, Vth Ed.*, Edagricole, Bologna.
24. Colby, S.R., Hill, E.R., Kitchen, L.M., Lym, R.G., McAvoy, W.J., Prasad, R. (1989): *Herbicide handbook, VIth edition.* Weed Science Society of America, Champaign, Illinois.
25. Canton, J.H., Linders, J.B.H.J., Luttik, R., Mensink, B.J.W.G., Panman, E., Van de Plassche, E.J., Sparenburg, P.M., Tuinstra, J. (1991): Catch-up operation on old pesticides: an integration. March 1991. Rijksinstituut voor Volksgezondheid en Milieuhygiene (R.I.V.M.), Holland.

26. Verschueren, K. (1983): *Handbook of environmental data on organic chemicals*. IInd Ed, Van Nostrand Reinhold Company, New York.
27. World Health Organisation (1987): Drinking-water quality: guidelines for selected herbicides. *Environmental Health Series*, no. 27. WHO, Regional Office for Europe, Copenhagen.
28. Trevisan, M., Capri, E., Ghebbioni, C. (1991): L'uso di indici per prevedere la contaminazione delle acque. *Acqua e Aria*, 863-874.
29. Muir, D.C.G. (1991): Dissipation and transformations in water and sediments. In: Grover, R., Cessna, A.J. (Eds.), *Environmental Chemistry of Herbicides*, CRC Press, Boston, Vol. II, 1-89.
30. Howard, P.H., Boethling, R.S., Jarvis, W.F., Meylan, W.M., Michalenko, E.M. (1991): *Handbook of environmental degradation rates*. Lewis Publishers Inc. Chelsea.
31. Vighi, M., Masoero Garlanda, M., Calamari, D. (1991): QSARs for toxicity of organophosphorus pesticides to *Daphnia* and honeybees. *Sci. Tot. Env.* 109/110: 605-622.

6. PURIFICATION OF GROUNDWATERS POLLUTED BY ATRAZINE

C. NURIZZO
Faculty of Engineering,
Milan Polytechnic,
Piazza Leonardo da Vinci 32,
20133 Milan, ITALY.

1. Introduction

There are three main technical options available for the removal of atrazine and its related compounds (e.g. simazine, bentazone, etc.) from groundwaters; these are:
(a) Adsorption onto activated carbon;
(b) Chemical oxidation (e.g. with ozone);
(c) Filtration through membranes (e.g. reverse osmosis).

Of these, the most widely used option for large scale treatment is activated carbon, although the other two options are being used on a local scale. However, there are many studies in progress at present, either as laboratory tests or pilot (semi-scale) and full-scale trials, to assess the practicality of these different types of treatment and the possibility of their further improvement.

This paper considers the technical details of these treatment methods and their financial implications.

2. Alternative Treatment Methods

2.1. ACTIVATED CARBON

2.1.1. Powdered Activated Carbon
Activated carbon (charcoal) has been used for many years, in Italy and elsewhere, for the removal of harmful organic compounds (i.e. micropollutants, including atrazine) from water. For surface waters abstracted for public supply, for which the treatment works include flocculation tanks for the removal of suspended solids, the activated carbon can be added as a powder (PAC - powdered activated carbon) with the flocculating chemicals and subsequently settled out. In this case, the only new installation required is a silo for holding the PAC and a dosing device; there are unlikely to be any problems caused for the subsequent rapid filtration. However, the residence time of the PAC in the flocculation/clarifying system must be sufficient for an equilibrium to be reached with the organic compounds; the duration necessary is usually in the range of 40 to 60 minutes. Also, the PAC should be thoroughly mixed with the water and mean velocity gradients (G)

L. Bergman and D.M. Pugh (eds.), Environmental Toxicology, Economics and Institutions, 141-150.

must be within the range of 50 to 70 s^{-1}.

2.1.2. Granulated Activated Carbon

Alternatively, the activated carbon can be used in the form of granules (GAC - granulated activated carbon). Where groundwater is being treated (and where there are no flocculation tanks required) the GAC is used in a rapid filtration system operating under pressure. These GAC filter beds (usually with a 'downflow' of water) are 2 to 3 metres high with a maximum diameter (standard) of 3.6 metres; deeper or shallower beds can be used but these are rarely outside the range of H = 1.2 to 6.0 metres. The optimum geometry of the filter bed is based on the specific hydraulic load (SHL - or average filtration velocity) which is usually in the range of 5 to 30 m^3 water/m^2 surface area (= m/h); typical beds need a nominal contact time (NCT) of at least ten minutes, which equates with a SHL of about 15 m/h. For readily adsorbed molecules such as herbicides, the NCT would be in the range of 10 to 20 minutes with a corresponding SHL of 6 to 12 m/h.

In operational use, the GAC filter becomes saturated with the micro-pollutants from the surface downwards. When the absorptive capacity is exhausted, the GAC has to be reactivated; this is usually by heat treatment which leads to a loss in granule weight of 10% to 12%. There are only a few such regeneration facilities in Europe. The time taken to reach saturation (the adsorption cycle) depends on the bed depth, the type of GAC used, on the SHL, and in particular on the type and concentration of micro-pollutants present in the water. In the case of atrazine, the results of laboratory simulations, which have been confirmed by operational experience over the past five years, show that an adsorption cycle of 18 to 20 months can be obtained if the groundwater concentration of micropollutants (Co) does not exceed 0.5 μg/l; for lower values of Co the cycles are longer (Figure 2). With such long cycles, the filter bed can become compacted and short circuiting of the flow occurs; back-flushing to loosen the bed has to be carried out usually every two weeks.

Due to the need to back-flush and to reactivate the GAC, it is usual to install at least two filters (even in small treatment plants at a well-head), so that continuous operation can be maintained.

An alternative use of GAC is in fluidised ('upward flow') beds. This avoids the problems of compaction and the need for back-flushing; also, the saturated GAC is heavier than the fresh granules, and these gravitate to the bed bottom where they can be removed at intervals for reactivation and replaced by new GAC.

As stated above, the water chemistry is an important factor in determining the efficiency of GAC filters. Figure 1 shows the specific throughputs for different herbicides as a function of bed depth, using a filter containing a specific GAC (F100) and with a SHL of 10 m/h [1]. Although the effect of increased levels of total organic carbon (TOC) in reducing the cycle times is mainly a problem for surface waters, groundwater TOC concentrations can frequently exceed 0.5 mg/l and may reach 1 mg/l. Figure 2 shows the effect of the TOC content of water on the cycle times achieved for three Cos (0.2, 0.5 and 1.0 μg/l), and the NCT was short at 10 minutes [2].

2.1.3. Theoretical Considerations

The adsorption capacity of activated carbons for micropollutants can be modelled in different ways. For very dilute solutions (e.g. such as groundwaters) the usual mathematical model is based on the Freundlich equation:

$$q = k.Ce^{1/n}$$

where q = X/M (the activated carbon efficiency expressed as g micropollutant adsorbed/g carbon.)

Ce = the equilibrium concentration of the micropollutant in the water (µg/l) at constant temperature.

k, n = constants which depend on the type of micropollutant molecule and the properties of the particular activated carbon used.

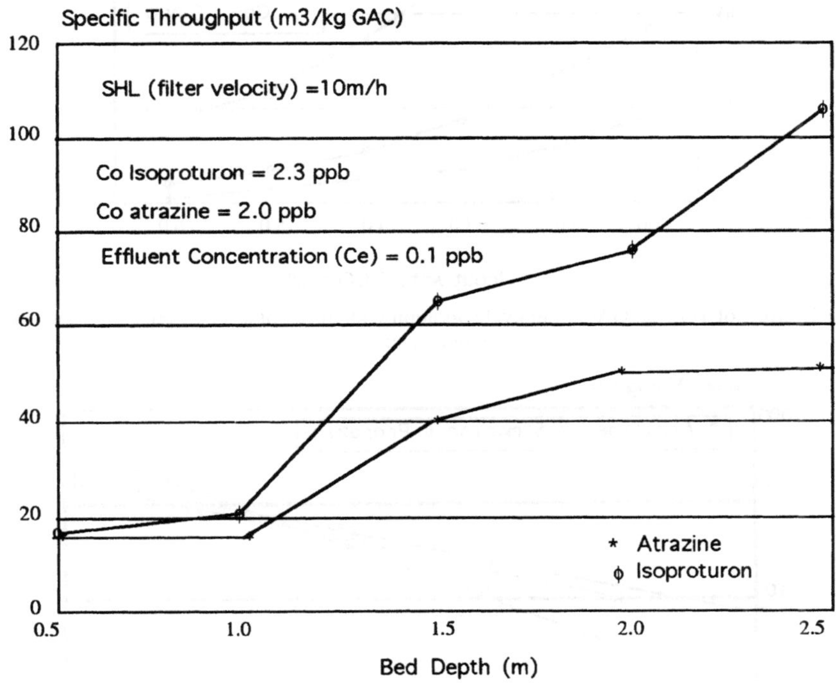

Figure 1. Specific f(H) throughputs of GAC filters.

Figure 3 shows linear transformations of the log-log Freundlich isotherms for a specific GAC (F200) for three triazine herbicides [3]. It is clear that there are no differences between the behaviour of atrazine and propazine; for simazine, the activated carbon efficiency (q) is higher at higher equilibrium concentrations (Ce) and somewhat less at lower values of Ce. Thus, it would be expected that this specific GAC will be less suitable for simazine removal especially if low residual concentrations are required, but it will be effective for the removal of the other triazines.

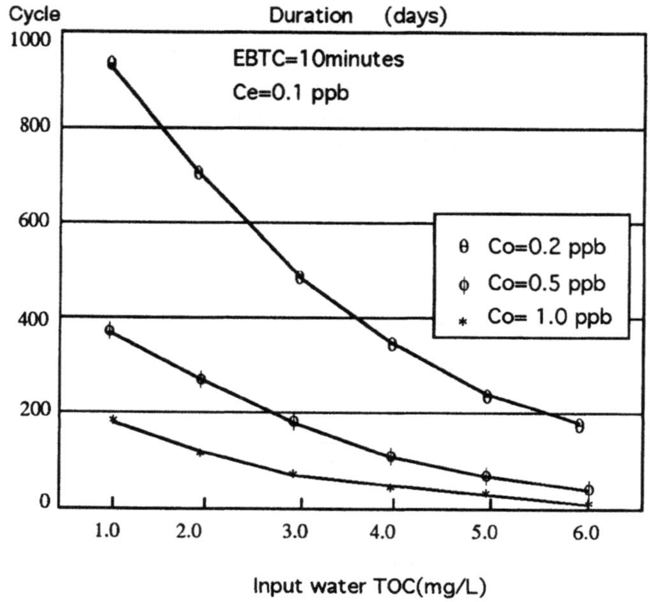

Figure 2. Effect of TOC on GAC filter cycle duration with different atrazine concentrations in the input water.

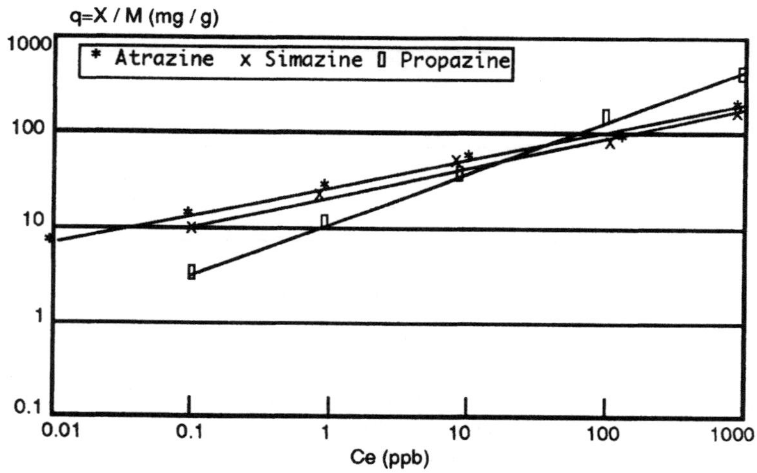

Figure 3. Freundlich isotherm for some herbicides on F200 carbon. A: atrazine; B: propazine; C: simazine.

2.1.4. Properties of GACs

These physical characteristics of importance in determining the performance characteristics of two GAC preparations commonly used in water treatment (F100 and F200) are compared in Table 1.

Table 1. Physical properties of two granular activated carbons (GAC) used in water treatment.

Main Parameters	Units	F100	F200
Total Surface	m^2/g	850-900	850-900
Bed Density*	kg/L	≈500	≈200
Particle Density^	kg/L	≈1.5	≈1.5
Nominal Diameter	mm	1.5-1.7	0.9-1.1
Effective Size	mm	0.8-1.0	0.6-0.7
Uniformity Coeff.	mm/mm	<1.9	<1.9

* backwashed and drained
^ wetted in water

GACs do not remove all classes of chemical micropollutants with equal efficiency. The general position for several such groups is compared in Table 2 [4].

Table 2. Adsorption capability of a typical activated carbon for various organic micropollutants in water.

Class and Types of Organics	Adsorption Capability
Aromatic solvents (benzene, toluene, etc.)	Good
Pesticides (aldrin, endrin, chlordane, dieldrin, etc.)	Good
Herbicides (atrazine, simazine, propazine, etc.)	Good, very good
Aldehydes and ketones	Poor

2.2. CHEMICAL OXIDATION

Some modern treatment works use powerful chemical oxidants, sometimes in combination, to remove atrazine; some of these processes are commercially registered under the trade names (Perozone, Peroxone and Ozocarb). Further studies are in progress, and the following paragraphs review the practical feasibility of these techniques.

2.2.1. Chlorine

Chlorine is used widely in water treatment works as a sterilant. Some experimental work has been carried out on its efficiency as an oxidant to remove atrazine, but at the dose commonly used (1 mg/l) the results showed that it was rather ineffective [5].

2.2.2. Ozone

Ozone shows a greater promise as an oxidant of atrazine, although the efficiency can be reduced by the presence of bicarbonate ions and organic acids which react with the active free radicals [2].

As with activated carbon, the contact time is important. Figure 4 shows the results that were recently obtained on a semi-scale plant at the Environment Section of the

D.I.I.A.R. (Milan Polytechnic), using low ozone doses to reduce the formation of by-products. In this experiment, the effect of different contact times was measured by treating the water with ozone and then dividing it into two samples. One sample was allowed to remain in a tightly closed bottle for 90 minutes before analysis of the atrazine content. The second was treated with sodium sulphite to reduce the residual ozone, so that the effective ozone contact time was less than 4 minutes. It is clear that the short contact times were insufficient to obtain a reasonable removal of atrazine by comparison with the longer periods.

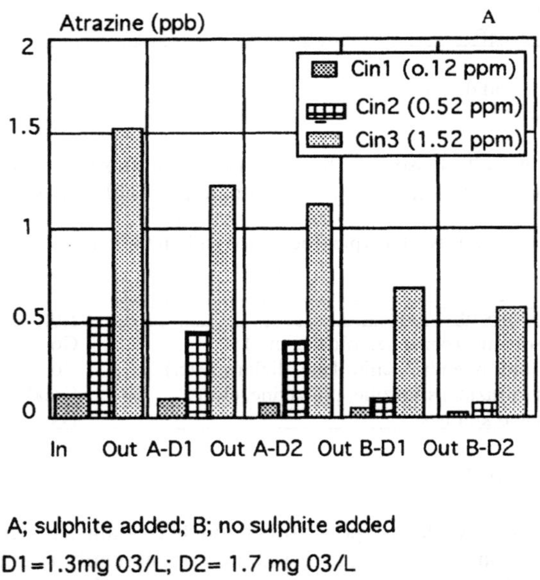

Figure 4. The effect of starting concentration of atrazine and the duration of exposure (A ≤ 5 min; B ≤ 90 min) to low dose ozone (1.3 or 1.7 mg/L) on atrazine depletion in water.

This experiment was later extended to include other herbicides. Simazine behaved in a similar way to atrazine, but the removal rates were better with molinate and very good for bentazone even with extremely short contact times of a few seconds.

2.2.3. Combination of Ozone and Other Treatment Methods

Other experiments have evaluated the combined effects of ozone and hydrogen peroxide, which give a higher free radical activity [2, 6, 7]. Figure 5 shows the results obtained for atrazine with medium to high doses of ozone alone and in combination with hydrogen peroxide; the contact time was less than five minutes [2].

However, one of the main problems of using strong oxidants such as ozone and hydrogen peroxide is the possibility that soluble by-products, perhaps in harmful concentrations may be formed in the water. For example, it was recently found that bromide in water treated with ozone was oxidised to the more toxic bromate; the proposed guideline value for bromate is very low at 0.5 µg/l, and the possibilities for removing this ion from the water are rather limited [8].

Other possible treatments include the combined use of ozone and UV light [9], with a secondary treatment using GAC to remove residual organic compounds. It is clear that further research into different treatment techniques is required, together with a careful appraisal of the associated problems that may arise.

2.3. FILTRATION THROUGH MEMBRANES

This technique (in practice confined at present to reverse osmosis) is possibly the least feasible of those considered here because of its high capital and running costs; also, the

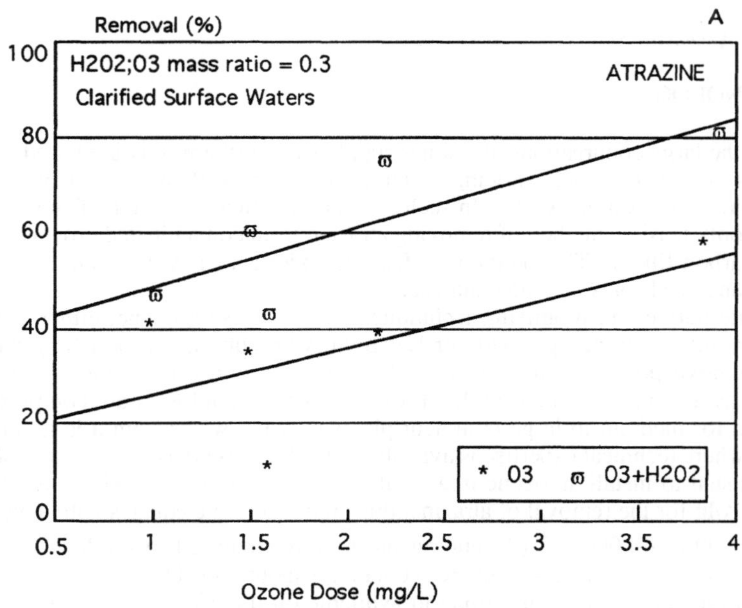

Figure 5. The effect of ozone concentration ± hydrogen peroxide on the removal of atrazine from water.

process removes dissolved salts which may not be desirable *per se* and these are concentrated in an effluent that requires subsequent disposal. There may, however, be occasions when the removal of salts such as nitrate, nitrite, sulphate, chloride, and other dissolved inorganic and organic compounds (including micropollutants such as atrazine) will be required and for which there are few alternative methods of removal available.

However, this situation may change with the development of 'nano filtration'; this modification of reverse osmosis employs a membrane that is more permeable to salts while still restraining organic molecules, and is less expensive to operate. In a recent study, the efficiency of removal of pesticides and other micropollutants by nano-filtration was assessed in a pilot-scale plant, using four different types of membrane [10]. The results were encouraging in that 80% to 98% of the atrazine was removed, depending on the membrane type. However, these results were obtained with atrazine solutions made up in water pre-treated by physical and chemical separation of suspended matter and by GAC filtration to remove organic material. Even so, severe fouling of the membrane was found to occur after a few days of operation. It is clear that this technique is promising , but more studies are required on the extent of the pre-treatment required and on possible improvements; also, the real operational costs (chemical energy, membrane durability, specialist staff, etc.) need to be evaluated.

3. General Assessment

3.1. TECHNOLOGY

Except for the large conurbations, the water supplies in Northern Italy are based mainly on small local authorities, each drawing water from a few wells and often only partially interconnected into local networks. In such a situation with a dispersion of small treatment works, the only feasible method of removing atrazine from contaminated groundwater is by activated carbon filters. This method is efficient under a variety of conditions; it can be automated and needs only routine maintenance.

Alternative and promising techniques are those using ozone, either alone or in conjunction with hydrogen peroxide or UV light, with subsequent polishing with GAC filters to remove possibly harmful break-down products; reverse osmosis using nano-filtration may also prove to be useful. However, these technologies are complex and are more suited for medium to large treatment plants drawing water from a large number of wells, and where technical expertise is available. It is for these reasons, for example, that it is most unusual to install an ozone treatment facility which has a production rate of less than 5 kg O_3/h; for the removal of atrazine, this production rate equates with a daily water flow of 30,000 to 40,000 m^3 (depending on the dose required to produce an effective concentration) which would serve an area containing about 100,000 inhabitants.

So far, there is little information available on the efficiency of these treatment methods for the removal of atrazine substitutes. Figure 6 shows a comparison between removal of atrazine and terbutylazine by ozone under identical operational conditions [11] and it is clear that the efficiency of such treatment is similar for both herbicides. Nevertheless, the first choice for the safe removal of organic micropollutants is GAC adsorption although more information is required on possible desorption and cycle duration when used for the removal of atrazine substitutes.

3.2. ECONOMICS OF WATER TREATMENT

The costs of construction and operating these various water treatment technologies vary considerably with differences in plant size and general layout, and in the initial concentrations of atrazine and other organic micropollutants in the water. However, the following information can be taken as an approximate indication in respect of GAC filtration.

3.2.1. Construction Costs

A recent investigation carried out by a Technical Study Committee of the Regione Lombardia [12], subsequently confirmed, indicated that the capital cost of installing a treatment plant at the well-head would be in the range of 1.5 - 1.8 million ITL for 1 m^3/h of nominal flow. Thus, a GAC plant to treat 50 m^3/h would cost up to 90 million ITL. Similar costs have been given for GAC filtration of pre-treated surface waters at large treatment plants.

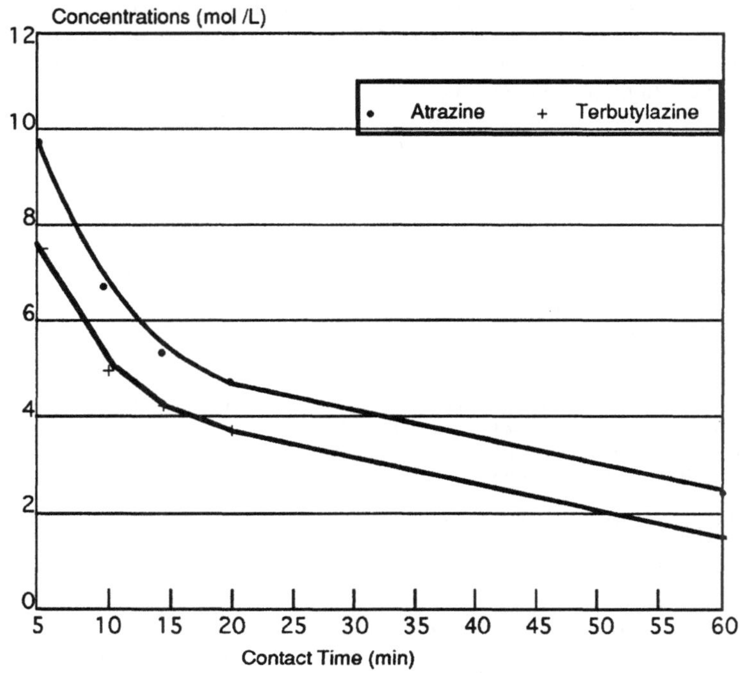

Figure 6. Efficiency of ozone treatment (dose D = 3.5 mg/l, pH 7) on the removal of atrazine and terbutylazine [11].

3.2.2. Operational Costs

The general costs that have been calculated are in the range of 50 to 70 ITL/m^3, depending on the initial atrazine concentration and the presence of other organic micropollutants, and the type of treatment plant. These figures include the costs of energy, manpower, reactivation of GAC on a 600 day cycle, and maintenance; also, depreciation costs based on a 12% annual discount for 12 years in the case of electrical and mechanical items and 25 years for civil construction works.

These operational costs for GAC may be reduced if the filters are used in conjunction with other treatment techniques (e.g. ozone and hydrogen peroxide) as described earlier, but the efficiency and acceptability of these methods is still under review.

3.2.3. Household Treatment

Where a single house or small community is served by its own water supply, a possible remedial technique to remove micropollutants is to attach a small treatment unit to the drinking water tap. These units may be simply replaceable GAC cartridges, or the more technically advanced small membrane filters with subsequent GAC filtration and UV sterilisation. Such units are suitable only for very low flows (usually not more than 50 litres/day) and are therefore suitable only for drinking water supply and not for general household use. The initial cost of an advanced unit is in the range of 1.5 to 2.0 million ITL, and the running cost (mainly depending on the approximately annual replacement of the membrane) is very high, the range being 15,000 to 18,000 ITL m^3.

4. References

1. Haist-Gulde B. *et al.* (1992) Removal of pesticides from raw waters, *Proc. IWSA Specialised Conference AQUATECH92, Water Supply*, 159-168.
2. Duguet J.P. *et al*. (1991) Removal of atrazine from Seine river by ozone-hydrogen peroxide combination in a full scale plant, *Proc. 10th Ozone Congress*, 301-312.
3. Chemviron Carbon (1990) *Technical Information,* Chemviron.
4. Faust S.D., Aly O.M. (1983) *Chemistry of Water Treatment*, Ann Arbor Sciences..
5. Cremisini C. *et al*. (1990) Valutazione degli effetti della disinfezione delle acque con ipoclorito sulle triazine ed altri erbicidi di interesse igienico-sanitario, *Acqua & Aria*, 7/8, 581-587.
6. Chramosta H. *et al* (1991) Effects of alkalinity and dissolved organic matter on the degradation of atrazine by O_3 and H_2O_2, *Proc. of 10th Ozone Congress*, 281-300
7. Bouillot P. *et al.* (1991) Efficiency of different possibilities of making use of ozone, alone or coupled, as regards the refractory organic carbon transformation in carbon biodegradable by bacteria. Personal communication.
8. Kruithof J.C., Schippers J.C. (1992) The formation and removal of bromate, Proc. *IWSA Specialised Conference AQUATECH92*, 121-127.
9. Duguet *et al.* (1989) - Efficacy of combined use of ozone/UV or ozone/hydrogen peroxide for water disinfection, *Proc. I.O.A. Symp. "Ozone and UV water treatment"*, 264-271.
10. Hofman H. *et al* (1992) - Removal of pesticides and other organic micropollutants with membrane filtration, *Proc. IWSA Specialised Conference AQUATECH92*, 101-111.
11. Brauch, H.J., Werner P. (1991) - Identification and characterisation of ozonization by-products of selected pesticides by chemical analysis and bio-test, *Proc. of the 10th Ozone Congress*, 165-178.
12. Comitato Tecnico Regionale (1992) - *Piano regionale di risanamento delle acque*, Regione Lombardia, Settore Ambiente ed Ecologia.

7. THE COSTS OF MEETING A DRINKING WATER QUALITY STANDARD: THE CASE OF ATRAZINE IN ITALY

TORE SÖDERQVIST[1]
Department of Economics
Stockholm School of Economics
Sweden

7.1. Introduction

In general, man-made contamination of groundwater is an unintentional outcome of the activities of those engaged in, for example, agricultural or industrial production, or of households' consumption activities. A fundamental problem is that these people whose activities cause the contamination are often not those that are subsequently affected by it. This means that the activities which cause contamination are not automatically restricted by the damage caused by them. Such contamination is an example of a *negative externality*: the behaviour of one group of people has a damaging impact on the well-being of a second group of people, but this impact is not taken into account by the first group.

The concept of externality is illustrated in Figure 7.1 for the particular case of pesticides which contaminate groundwater. In the figure's *society* box, we find two *private decisions* boxes; one includes producers and users of pesticides, say firms and farmers. The exchange of pesticides for money takes place in the pesticide market. However, the behaviour of firms and farmers is not automatically influenced by the externalities caused by them. The *externalities* box shows how pesticides are distributed to recipient compartments in the environment (e.g. soil, air, surface water, etc.). If the pesticides have a sufficient mobility and persistence, a proportion will eventually reach groundwater. Such groundwater may then be used as a drinking water source and abstracted by public or private waterworks, or by household wells, as illustrated in the second private decisions box. Consumers of drinking water can decide, at least in principle, what quality of drinking water they would like to consume and exchange drinking water for money in such a market.

1 The author is indebted to, among others, Lars Bergman, Michael Faure, Enzo Funari, Per-Olov Johansson, Richard Lloyd, Constantino Nurizzo, D. Michael Pugh, Timothy M. Swanson, Marco Vighi, and Giuseppe Zanin for data, help and comments. The usual disclaimer applies. Financial support from the European Science Foundation and the Swedish Council for Planning and Coordination of Research is gratefully acknowledged. The author is presently visiting The Beijer International Institute of Ecological Economics, Royal Swedish Academy of Sciences, Box 50005, S-104 05 Stockholm, Sweden.

L. Bergman and D.M. Pugh (eds.), Environmental Toxicology, Economics and Institutions, 151-171.

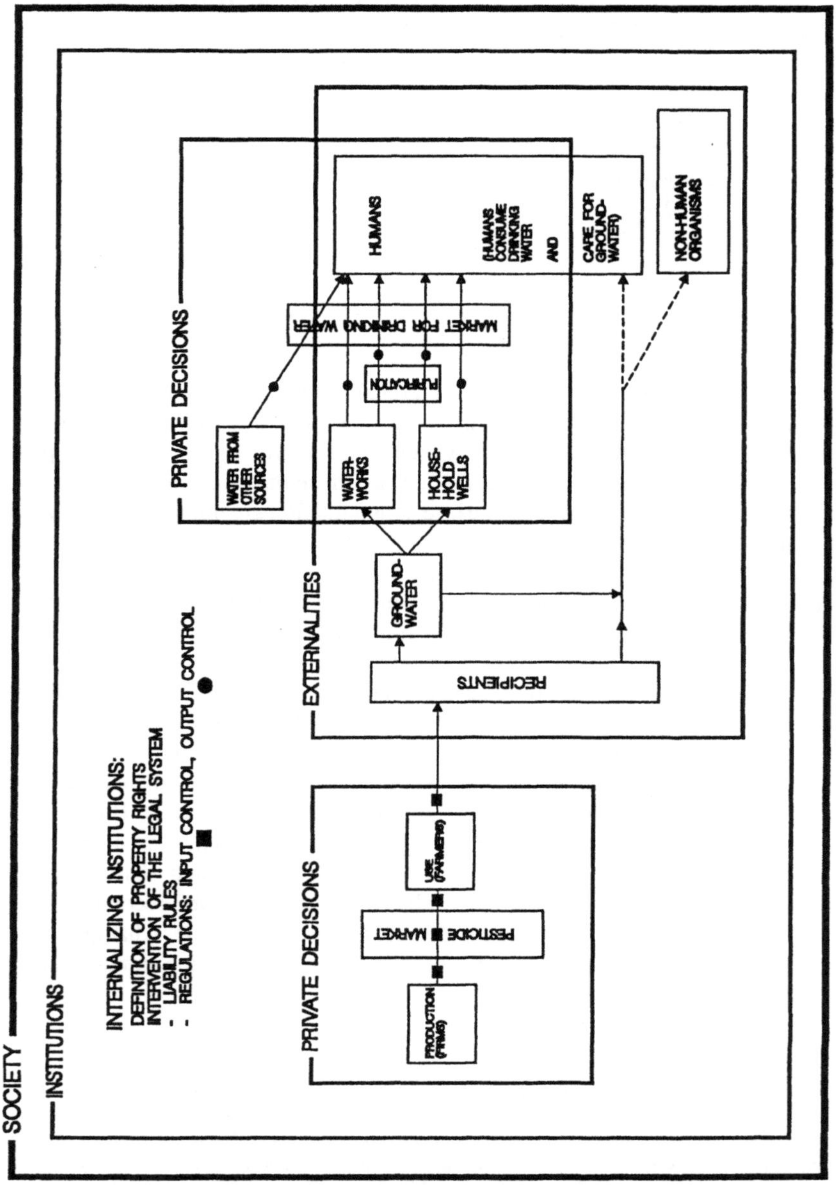

FIGURE 7.1. A rough flow chart for a pesticide which contaminates groundwater.

The boxes for private decisions and externalities are integrated by introducing *institutions* (*"the rules of the game in a society"* (North (1990, p. 3)) which constrain the private decisions in such a way that the decisions have to take the externalities into account, i.e. the externalities are *internalized*. In the Figure, this is illustrated by an institutions box which encompasses the private decisions and externalities boxes. Ideally, the constraints imposed should be based on an overall analysis, in which the *social* benefits and costs of the production and use of a pesticide are estimated. In such a cost-benefit analysis, the externalities box is taken into account. For such an analysis to be successful, there should be, *inter alia*, sufficient scientific knowledge of how and to what extent different pesticides can enter groundwater and what damage they then cause. This analysis may determine the *socially efficient* quantity of the pesticide (which may be zero or above zero), and institutions enforcing the socially desirable situation should then be constructed. It has been shown that, in theory, a simple definition of property rights may suffice for such an enforcement. In reality, however, a more active intervention by the legal system is likely to be necessary through the definition of liability rules and/or regulations.[2]

A different situation arises if a regulation, e.g. a drinking water standard, has been set and it is then discovered that drinking water does not meet this requirement. The desirable action that should be taken to improve the water quality is that which enables the standard to be met at the lowest cost, i.e. the *cost-effective* action. For example, in the case of pesticides, the costs of reducing pesticide use may be compared with water purification costs; both these actions may have the potential to enable the standard to be met.

This chapter follows the route of a cost-effectiveness analysis and focuses on the costs for meeting a drinking water standard set by the European Community (EC), now the European Union. The case study chosen is the presence of the herbicide atrazine in groundwater in northern Italy, and the Italian decision to ban the use of this particular herbicide in 1990. Section 2 of this chapter gives the background to this case study. In Section 3, the increased costs incurred by the farmers because of the atrazine ban are estimated by two different approaches. Section 4 is devoted to water purification costs. Section 5 completes the chapter with a discussion which includes some conclusions. This last section also contains some remarks on the very limited knowledge of the benefits derived from reduced atrazine levels in groundwater. This lack of knowledge meant that a determination of the socially efficient concentration of atrazine in groundwater could not be undertaken.

Before we consider the chosen case study, the important economic concepts that were touched upon in the preceding paragraphs deserve some comment. It is beyond the scope of this chapter to examine these concepts in detail, but the relevant information is given in other publications. Johansson (1991) gives an introduction to modern welfare economics, i.e. the branch of economics devoted to normative issues, and Johansson (1993) considers an important area within applied welfare economics, that is the cost-benefit analysis of

---

2 Another and usually very powerful type of institution is restrictions set by ethics, i.e. moral principles.

environmental change. Other books on welfare economics include Boadway and Bruce (1984) and Just *et al.* (1982). Söderqvist *et al.* (in preparation) provide a very basic introduction to groundwater contamination as an economic problem. They also consider the issue of what institutions could be established in order to enforce a socially desirable situation, and then focus on the option of using economic instruments, e.g. charges and permit markets. Another chapter in this book, Faure (1994), is devoted to institutional aspects. Among other things, Faure discusses the conditions under which either liability rules or regulations would be the preferred type of intervention. He also emphasizes the important distinction between input control and output control in e.g. the case of pesticides (cf. Figure 7.1). For descriptions of economic instruments, the reader is referred to Opschoor (1991) for the particular case of persistent pollutants, and to Baumol and Oates (1988), Bohm and Russell (1985), Opschoor and Vos (1989), and Tietenberg (1990, 1992) for a general background.

7.2. The Case of Atrazine in Italy

Atrazine is a herbicide which belongs to the chemical family of triazines. In Italy, atrazine has been used extensively in maize cultivation since the early 1960s. Vighi and Zanin (1994) estimate that at least 90 per cent of the area used for maize production was treated with atrazine, either alone or in combination with other herbicides. One of the advantages of using atrazine in maize cultivation is the flexibility in the timing of the application: it can be used at pre-sowing, at pre-emergence or post-emergence of weeds, and at any phase of maize growth. Unfortunately, it has also leaching properties, which make it likely that some of the atrazine used will eventually reach groundwater, especially in areas with permeable soils.

The EC Council Directive on Drinking Water Quality of July 15, 1980, set the maximum admissible concentration for individual pesticides in drinking water at 0.1 μg/L.[3] This standard thus applies to atrazine. It is considerably more restrictive than the toxicologically based 2 μg/L guideline set for this herbicide by the World Health Organization. The EC Council Directive was implemented into Italian law in 1985. In some areas in the Po Valley in northern Italy, where maize cultivation is extensive and groundwater the dominant source of drinking water, chemical monitoring of drinking water revealed that concentrations of atrazine exceeded the EC standard. As a consequence, enforcement of the EC Directive in respect of atrazine began in 1986 with local restrictions on the use of drinking water supplies where the standard was exceeded. In March 1990, the Italian Ministry of Health decided to introduce a nationwide ban on the sale and use of atrazine.

---

[3] See Faure (1994) for a detailed description and analysis of the EC Council Directive, and of the Italian implementation and enforcement of the Directive.

In principle, the introduction of such a ban is only one of the possible ways of enforcing the EC Directive. If the preferred option is that atrazine should not reach groundwater, charges or permit markets are examples of other instruments that could be used to restrict its use. However, as illustrated by a simulation model constructed by Zanin *et al.* (1993), under certain weather conditions (e.g. a rainy May or June), the groundwater in areas with permeable soils is likely to become contaminated even if only small amounts of atrazine are used. This fact provides a crude, but none the less valid, argument for introducing a ban: only zero use is certain to ensure that the EC standard for the banned pesticide will be met sooner or later. However, if atrazine can be tolerated in groundwater *per se*, but not in drinking water, purification of the abstracted water can be used as a way of meeting the EC standard. Water purification may thus be used as an alternative to a ban if e.g. no environmental aspects are taken into account, but only drinking water quality. Which of these measures, a ban or water purification, is in this particular case the cost-effective one, i.e. which measure ensures that the EC standard is met at the lowest cost? This question is approached in the next Section by estimating the farmers' increased costs due to the ban on atrazine, and examine water purification costs in Section 4.

7.3. Costs of the Ban

Among the many types of possible costs incurred because of the introduction of the ban, those borne by the farmers who cultivate maize are in particular likely to be considerable. Atrazine was the herbicide most commonly used in maize cultivation, which shows that it had a considerably superior performance over that of other herbicides. All other possible costs will be disregarded in this chapter, although some of them might not be inconsiderable. For example, the producers of atrazine and the users of atrazine for purposes other than maize cultivation, and the enforcement of the ban, may incur additional costs.

Two different approaches will be used to estimate the increased costs incurred by maize farmers: first, a *micro-based* approach which focuses on the substitution pattern of farmers in a representative area in northern Italy, and then a *macro-based* approach based on the aggregate demand for atrazine.

7.3.1. A MICRO-BASED APPROACH[4]

Let us assume that farmers are profit maximizers and produce maize (m) by the use of weed control based on atrazine (a), a vector of other herbicides ($\boldsymbol{h}$) and a vector of all inputs other than a and $\boldsymbol{h}$ ($\boldsymbol{x}$). This implies a production function $m=f(a,\boldsymbol{h},\boldsymbol{x})$. If p_m, p_a, $\boldsymbol{p_h}$ and $\boldsymbol{p_x}$

---

[4] The estimation of costs which follows this approach relies heavily on data collected by Professor Giuseppe Zanin, University of Padova.

denote the prices of maize, weed control based on atrazine, other herbicides (a vector) and all inputs other than a and $\boldsymbol{h}$ (a vector), the profit maximization problem for a farmer could be written as:

$$max_{a,\mathbf{h},\mathbf{x}}\ [p_m f(a,\boldsymbol{h},\boldsymbol{x}) - p_a a - \boldsymbol{p_h h} - \boldsymbol{p_x x}] \tag{7.1}$$

The solution to this problem is given by the profit function $\Pi(p_m,p_a,\boldsymbol{p_h},\boldsymbol{p_x})$, which gives the maximum profits (π) (e.g. see Varian (1992) for details), i.e.:

$$\pi = \Pi(p_m,p_a,\boldsymbol{p_h},\boldsymbol{p_x}) \tag{7.2}$$

Now let 0 superscript denote the situation before the atrazine ban and 1 the post-ban situation. For farmers who used a weed control based on atrazine, we have

$$\pi^0 = p_m^0 m^0 - p_a^0 a^0 - \boldsymbol{p_x^0 x^0} \tag{7.3}$$

Because of the atrazine ban, $a^1=0$ and the farmers had to turn to other herbicides instead, i.e.:

$$\pi^1 = p_m^1 m^1 - \boldsymbol{p_h^1 h^1} - \boldsymbol{p_x^1 x^1} \tag{7.4}$$

Even if the weed control efficiency of $\boldsymbol{h}^1$ is equal to that of a^0, the atrazine ban might very well have induced farmers to change their demand for $\boldsymbol{x}$, so there is no reason to believe that $m^1=m^0$. Also, the possible changes in aggregate demand and total production might have influenced market prices, but it is not obvious how any such effects due solely to the atrazine ban can be identified. I will adopt a partial view here and only study the cost change $\boldsymbol{p_h^1 h^1}-p_a^0 a^0$.

Detailed data on maize farmers' short-term substitution behaviour following the atrazine ban are available for an area called Alto vicentino. This area is a part of the plain north of Vicenza in the Veneto region. Since the end of the 1960s, 15-25 per cent of the area's 27,200 hectares has been used for maize cultivation. Table 7.1 shows estimates of annual herbicide costs for maize farmers when they were free to use atrazine. It is evident that weed control was based primarily on atrazine, either applied alone or in combination with other herbicides. When the direct herbicide costs and the costs of lost production caused by a weed kill rate below 100 per cent are taken into account, annual total costs can be estimated to about MITL (millions of Italian lire) 641 or ITL 112,400/ha.[5] If the farmers in the area of Alto vicentino are considered to be representative of those elsewhere in Italy, the total Italian herbicide cost for maize cultivation in 1989 was calculated to be

---

5 All monetary values are stated in 1990 prices, if not otherwise stated. ITL 1 corresponds to about USD 0.00075.

TABLE 7.1. Yearly herbicide costs for maize cultivating farmers in Alto vicentino before the atrazine ban

Treatment	Usage share (%)	Treated hectares	Direct herbicide costs/ha (ITL)	Kill rate (%)	Costs of lost production/ha (ITL)	Total costs (MITL)
az+ac	40	2,280	58,000	95	65,300	281.124
az+mc	20	1,140	68,000	95	65,300	151.962
az+osarh	5	285	90,000	94	78,100	47.909
az	30	1,710	16,000	95	65,300	139.023
az+sz	5	285	20,000	96	52,700	20.720
Total	100	5,700				640.738

Total cost per ha = 640.738/5700 = ITL 112,410

Legend: az = atrazine, ac = alaclore, mc = metolaclore, osarh = other soil applied residual herbicides which kill primarily grass weeds, sz = simazine.

TABLE 7.2. Yearly herbicide costs for maize cultivating farmers in Alto vicentino after the atrazine ban

Treatment	Usage share (%)	Treated hectares	Direct herbicide costs/ha (ITL)	Costs of lost production/ha (ITL)	Total costs (MITL)
sa+sarh	74.0	4,372	90,560	95,100	811.706
sa+sarh+apt	11.5	672	156,670	44,000	134.850
mix of sarh	8.5	504	129,400	69,450	100.220
mix of sarh+apt	1.0	57	199,000	13,500	12.112
only post	5.0	295	32,200	187,200	64.723
Total	100.0	5,900			1123.611

Total cost per ha = 1123.611/5900 = ITL 190,443

Legend: sa = substitutes for atrazine (e.g. terbuthylazine, linuron and cyanazine), sarh = soil applied residual herbicides which kill primarily grass weeds, apt = additional post-emergence treatments, post = post-emergence treatments.

112,400 · 778,064 · 0.9 = 112,400 · 700,258 = MITL 78,709, where 778,064 is the total area in ha cultivated with maize in Italy in 1989 and 0.9 reflects the fact mentioned above that weed control based on atrazine was used in at least 90 per cent of the total area.

Table 7.2 shows the farmers' short-term herbicide substitutions which occurred because of the atrazine ban. The main substitutes for atrazine are terbuthylazine and linuron, used on about 68 and 19 per cent of the hectares treated respectively, and always in combination with other herbicides. The Table shows that the annual post-ban herbicide costs can be estimated to be MITL 1,124 or ITL 190,400/ha. This corresponds to a cost increase of 1,124 - 641 = MITL 483 or 190,400 - 112,400 = ITL 78,000/ha. In 1990, the total Italian area cultivated with maize decreased to 742,285 ha, but increased to 835,362 ha in 1991. It is not known whether the atrazine ban has influenced farmers' decisions to cultivate maize, so let us simply suppose that the farmers who cultivated 700,258 hectares with the aid of atrazine in 1989 continued to do so in 1990 and that they complied with the ban. The annual herbicide costs for treating these hectares then increased to 190,400 · 700,258 = MITL 133,329, i.e. an increase in annual costs by 133,329 - 78,709 = MITL 54,620.

It is clear that the calculation of this cost increase figure is based on a considerable amount of simplification. Would the results be changed significantly if, for example, an area other than Alto vicentino was selected as the representative area? Some relevant information was provided by data on the short-term substitution behaviour of farmers in another area of maize cultivation near Brescia in the Lombardia region, where 45,000 ha were treated with herbicides. Terbuthylazine was also the main substitute for atrazine in this area. The increase in the farmers' annual herbicide costs can be estimated to be about ITL 98,000/ha, as compared to ITL 78,000/ha for Alto vicentino. If the former cost increase is used, the national cost figure would be MITL 68,625, i.e. MITL 14,005 larger than the corresponding figure based on the Alto vicentino data. Another question is whether similar results would be obtained using another type of data and another method of estimation. Let us therefore now turn to the macro-based approach and its results.

7.3.2. A MACRO-BASED APPROACH

Let us return to the profit function $\Pi(p_m,p_a,p_h,p_x)$. By Hotelling's lemma (e.g. see Varian (1992)), it is known that $\partial\Pi(.)/\partial p_a = -a(p_m,p_a,p_h,p_x)$, where $a(.)$ is a farmer's demand function for weed control based on atrazine. In principle, the change in profits due to the atrazine ban can now be calculated as:

$$\Delta\pi = \Pi(p^0_m,p^\infty_a,p^0_h,p^0_x) - \Pi(p^0_m,p^0_a,p^0_h,p^0_x) = \int_{p^0_a}^{\infty}[\partial\Pi(p^0_m,p_a,p^0_h,p^0_x)/\partial p_a]dp_a = \int_{p^0_a}^{\infty}-a(p^0_m,p_a,p^0_h,p^0_x)dp_a \tag{7.5}$$

where $p^\infty_a \geq p^*_a$ and $a=0$ when $p_a \geq p^*_a$.

$\Delta\pi$ is also illustrated by Figures 7.2a and 7.2b which show two possible shapes of the demand curve for atrazine. A "choke price" which causes a farmer to change discontinuously to other herbicides may very well exist (Figure 7.2b). Since atrazine cannot be described as an essential production factor (short-term substitution has in fact taken place), the expression for $\Delta\pi$ only applies if it can be assumed that all prices other than p_a are constants. I will rely on such a partial framework here and use data on the Italian market for atrazine to estimate an aggregate demand function for atrazine.

Atrazine was registered in Italy in 1963. It is probable that the use of atrazine increased steadily during the 1960s and at the beginning of the 1970s. Figure 7.3 shows the annual quantity sold of commercial products containing atrazine in Italy for the period 1973-1989. About 50 per cent of each commercial product consists of the active ingredient atrazine. The price of herbicides containing atrazine is that for the product Gesaprim L 1 kg. Price data are available for the period 1973-1987 by courtesy of Ciba-Geigy. The vertical axis of Figure 7.4 gives the real price paid by farmers for Gesaprim L. Current prices were adjusted to 1987 prices by the GDP deflator. It is clear that the real price decreased considerably during this period. This was a general trend not only for herbicides, but also for pesticides taken as a whole. Price and quantity data are combined in Figure 7.5.

Because of lack of data, the supply side of the Italian market for atrazine will be disregarded here. To use market data on prices and quantities for the estimation of a demand function without due regard to the possible identification problem is in general a very dubious activity. It may only be justified when there are reasons to believe that the supply curve is more prone to shift than the demand curve. It is not self-evident that this is the case here, and an important area for future research is thus to obtain more information on the supply side. With these words of caution in mind, we now turn to the estimation of the demand function.

Apart from the price of atrazine, the price of maize, of herbicides other than atrazine and of all other inputs appear as arguments in the demand function for atrazine above. Unfortunately, it has not been possible so far to obtain comparable price data on the common substitutes for atrazine. Instead, four aggregate price indices for the means of agricultural production will be used in the empirical version of the demand function for atrazine. Figure 7.6 shows real price indices for maize, herbicides, input I (total), input II (machinery) and input II (buildings).[6]

In order to keep the analysis at a basic level, let us work with two simple and commonly used functional forms for demand functions, viz. the "linear" and the "log-linear" forms:

---

6 "Input I" is defined as goods and services currently consumed in agriculture, e.g. feeding-stuffs, fertilizers, energy and lubricants, and "input II" as goods and services contributing to agricultural investment. The influence of the price of atrazine on the price index for herbicides, and the influence of the price index of herbicides on the price index for input I (total) have not been eliminated.

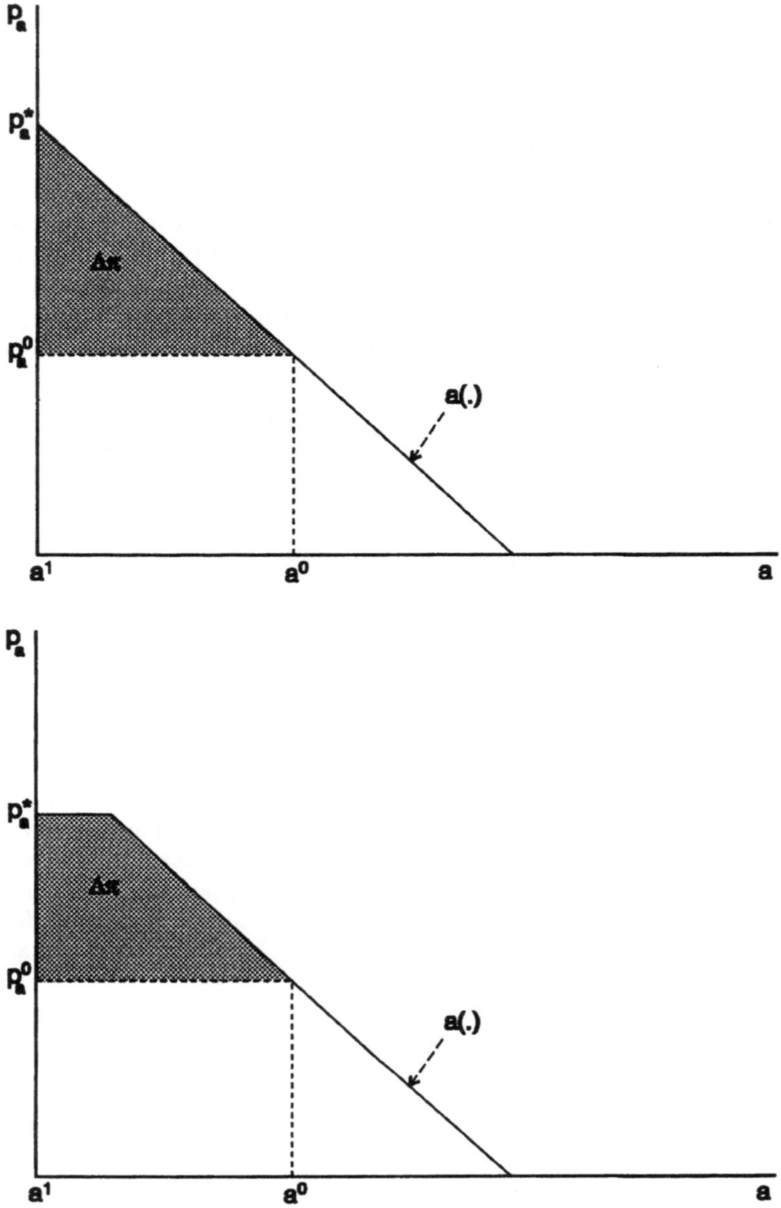

FIGURE 7.2a and 7.2b. Two possible shapes of the demand curve for atrazine.

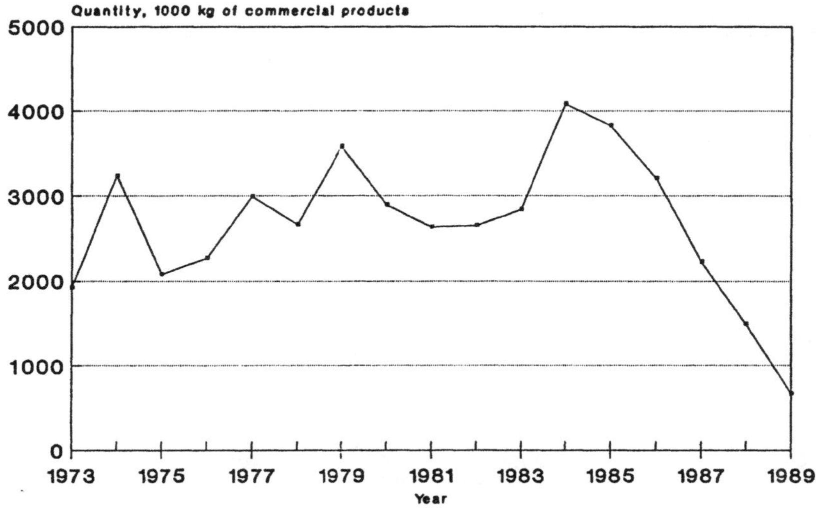

Sources:
Istituto Centrale di Statistica, Rome.
Lines and Tamaro [1990].

FIGURE 7.3. Quantity of atrazine sold in Italy.

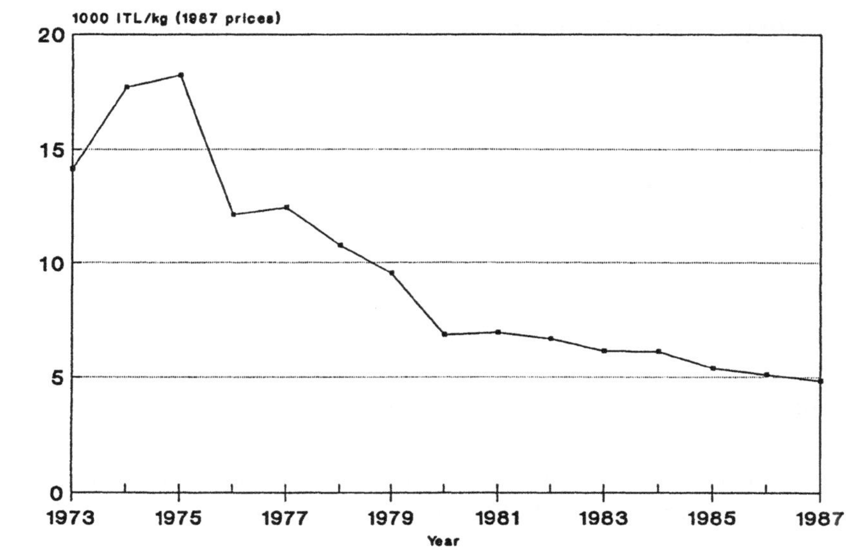

Source: Ciba-Geigy.

FIGURE 7.4. Real price of atrazine (Gesaprim L 1 kg).

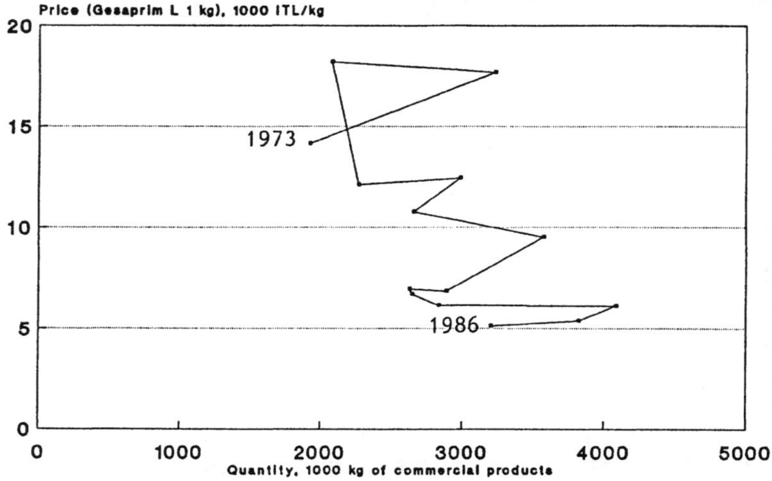

Sources:
Istituto Centrale di Statistica, Rome.
Ciba-Geigy, Lines and Tamaro [1990].

FIGURE 7.5. Price and quantity sold of atrazine.

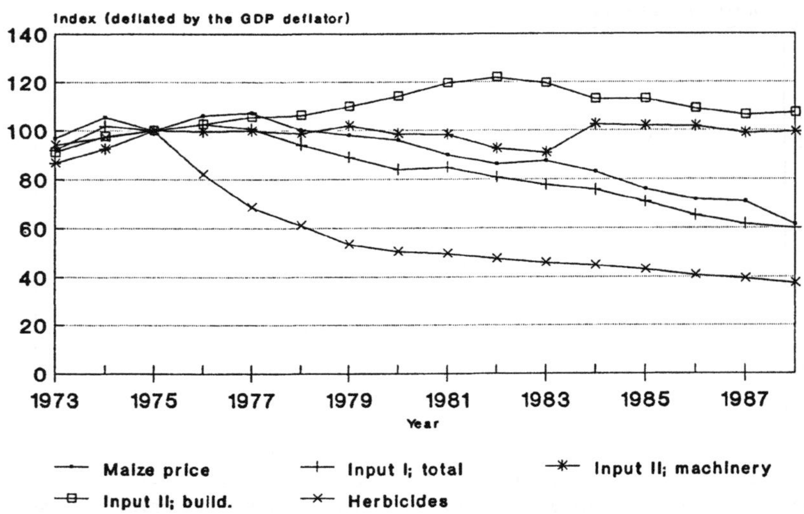

Source: Eurostat.

FIGURE 7.6. Deflated price indices (1975=100), Italy, for selected variables.

$$a_t = \alpha_0 + \alpha_1 p_{at} + \alpha_2 p_{mt} + \alpha_3 p_{ht} + \alpha_4 p_{it} + \alpha_5 p_{kt} + \alpha_6 p_{bt} + \epsilon_t \quad (7.6)$$

$$\ln a_t = \beta_0 + \beta_1 \ln p_{at} + \beta_2 \ln p_{mt} + \beta_3 \ln p_{ht} + \beta_4 \ln p_{it} + \beta_5 \ln p_{kt} + \beta_6 \ln p_{bt} + \eta_t \quad (7.7)$$

where a_t is the quantity of commercial products containing atrazine sold in year t in kg; p_{at} is the real price of atrazine (Gesaprim L) in year t in 1,000 ITL/kg; p_{mt}, p_{ht}, p_{it}, p_{kt}, p_{bt} are the real price indices of maize, herbicides, input I (total), input II (machinery) and input II (buildings), respectively, in year t; ϵ_t and η_t are stochastic error terms; and $\alpha_0,\ldots,\alpha_6,\beta_0,\ldots,\beta_6$ are coefficients to be estimated.

When using ordinary least-squares (OLS) estimators and observations for the years when there were no restrictions on atrazine use (1973-1986), inclusion of the data for 1974, characterized by a large and unique increase in both the real price and the quantity sold of atrazine, caused confusing results. Since there was no ready explanation for this, it was

TABLE 7.3a. OLS estimation results for the demand function $a_t = \alpha_0+\alpha_1 p_{at}+\alpha_2 p_{mt}+\alpha_5 p_{kt}+\epsilon_t$, t=1973,1975,...,1986

Coefficient	Estimate	t value	Level of significance
α_0	- 2613300	- 0.97	0.36
α_1	- 84318	- 1.94	0.08
α_2	- 5238	- 0.33	0.75
α_5	69145	2.87	0.02

Adjusted $R^2 = 0.617$ F = 7.46 Significance level of F = 0.01
Durbin-Watson statistic = 1.90 (the hypothesis of first-order autocorrelation can be rejected at the 5 per cent level of significance)

TABLE 7.3b. OLS estimation results for the demand function $a_t = \alpha_0+\alpha_1 p_{at}+\alpha_5 p_{kt}+\alpha_6 p_{bt}+\epsilon_t$, t=1973,1975,...,1986

Coefficient	Estimate	t value	Level of significance
α_0	- 1350828	- 0.38	0.72
α_1	- 117455	- 2.52	0.03
α_5	69243	2.91	0.02
α_6	- 13216	- 0.61	0.56

Adjusted $R^2 = 0.628$ F = 7.76 Significance level of F = 0.01
Durbin-Watson statistic = 2.02 (the hypothesis of first-order autocorrelation can be rejected at the 5 per cent level of significance)

TABLE 7.4a. OLS estimation results for the demand function $\ell na_t = \beta_0 + \beta_1 \ell np_{at} + \beta_2 \ell np_{mt} + \beta_5 \ell np_{kt} + \eta_t$, t=1973,1975,...,1986

Coefficient	Estimate	t value	Level of significance
β_0	4.556	- 1.11	0.30
β_1	- 0.3979	- 2.39	0.04
β_2	0.2369	0.43	0.68
β_5	2.1999	2.92	0.02

Adjusted $R^2 = 0.658$ $F = 8.69$ Significance level of F = 0.01
Durbin-Watson statistic = 1.95 (the hypothesis of first-order autocorrelation can be rejected at the 5 per cent level of significance)

TABLE 7.4b. OLS estimation results for the demand function $\ell na_t = \beta_0 + \beta_1 \ell np_{at} + \beta_5 \ell np_{kt} + \beta_6 \ell np_{bt} + \eta_t$, t=1973,1975,...,1986

Coefficient	Estimate	t value	Level of significance
β_0	5.915	1.18	0.27
β_1	- 0.3525	- 2.36	0.04
β_5	2.2102	2.90	0.02
β_6	- 0.0923	- 0.12	0.90

Adjusted $R^2 = 0.651$ $F = 8.47$ Significance level of F = 0.01
Durbin-Watson statistic = 1.91 (the hypothesis of first-order autocorrelation can be rejected at the 5 per cent level of significance)

decided to exclude this observation from the analysis. It was also found that there was a strong correlation between p_{at} and p_{ht} (r=0.96), and p_{at} and p_{it} (r=0.86). As a consequence, p_{ht} and p_{it} were excluded from the analysis, since p_{at} is a crucial variable. The regressions which included all the remaining variables p_{at}, p_{mt}, p_{kt} and p_{bt} showed statistically weak results, e.g. the coefficient estimate for p_{at} was not significantly different from zero at a 10 per cent level of significance. Instead, two alternative regression equations which include three price variables were used for further analysis. Results obtained from these equations are found in Tables 7.3a-3b (the linear forms) and 7.4a-4b (the log-linear forms).

The coefficient estimates for p_{at} have an expected sign. p_{kt} seems to have a surprisingly strong explanatory power and may very well be positively correlated with some other variable of importance for atrazine use, but which has not been taken into account here. For mean values for p_{at} (9.263) and a_t (2,894,044), the linear specifications give price elasticities of demand of -0.27 and -0.38, respectively. For the log-linear specifications, the estimates of β_1 give the elasticity directly as -0.40 and -0.35, respectively. For herbicides as a whole, an inelastic demand is not a surprising finding (cf. Dubgaard (1987) and Fogelfors *et al.* (1992) and see also the discussion in Andréasson-Gren (1992)), but for a single herbicide, an inelastic demand is reasonable only if there are no close substitute products. This is the case for atrazine, at least for alternative pre-emergence herbicides (Vighi and Zanin (1994)).

Let us now calculate the integral of Equation (7.5) for the two estimated linear demand functions

$$a_t = -2613300 - 84318p_{at} - 5238p_{mt} + 69145p_{kt} \tag{7.8}$$

$$a_t = -1350828 - 117455p_{at} + 69243p_{kt} - 13216p_{bt} \tag{7.9}$$

p^0_{at} is given by the most recent available value of p_{at}, i.e. that of 1987, which is equal to 4.842. Given index values for 1988 for p^0_{mt} (61.5), p^0_{kt} (99.4) and p^0_{bt} (107.2), p^*_{at} is equal to 46.699 in the case of Equation (7.8) and 35.036 when Equation (7.9) is used. This gives rise to the following two integrals:

$$-\int_{4.842}^{46.699}(-2613300 - 84318p_{at} - 322137 + 6873013)dp_{at} = MITL\ -73863 \tag{7.10}$$

$$-\int_{4.842}^{35.036}(-1350828 - 117455p_{at} + 6882754 - 1416755)dp_{at} = MITL\ -53541 \tag{7.11}$$

Even if these estimates of increased annual costs are converted to 1990 prices, they are still similar to those of the micro-based approach, which suggests that the increased annual cost for maize farmers due to the atrazine ban fall within the range MITL 50,000-100,000. Note, however, that the results of the macro-based approach is dependent on the choice of specification of the demand function; the integrals which correspond to Equations (7.10) and (7.11) in the log-linear case are divergent and are thus not realistic. Note also that both methods of estimation are partial and only concern farmers' short-term substitution behaviour. Again, it must be stressed that this increased cost for maize farmers is only a part of total costs due to the atrazine ban, though it is likely to be by far the largest part.

7.4. Costs of Water Purification[7]

The EC Directive mentioned in the introduction is concerned only with drinking water quality, i.e. it is not an environmental directive which applies to general groundwater or

---

7 The estimation of costs in this section relies heavily on data collected by Professor Constantino Nurizzo, Milan Polytechnic, and Professor Giuseppe Zanin, University of Padova.

surface water quality. This means that drinking water purification could be an alternative to an atrazine ban in order to meet the pesticide standard set in this EC Directive. However, if environmental aspects were to be taken into account, there would clearly be important differences between the use of an input control (e.g. the atrazine ban) and an output control (e.g. drinking water purification), in achieving this wider objective.

If an aquifer used for a public drinking water supply is contaminated by atrazine, several different measures can be taken in order to meet the EC standard. If the water from only one or a few wells contains an excess of atrazine, these can be closed temporarily and extra water can be taken from other wells not affected by the contamination. An alternative may be to dilute the contaminated water, i.e. to mix the water from the affected wells with water from other wells. Thus, careful monitoring and closure of affected wells, or possibly mixing water from different wells, are often sufficient measures for ensuring that the drinking water supplied meets the EC standard. However, if the contamination is severe and affects a large part of the aquifer, chemical purification of the water will be needed.

For individual households with their own private wells for family use, new problems are encountered. Monitoring becomes a less straightforward and relatively costly activity and the closure of a single private well has considerable complications for that affected household, who then has to obtain water from some other source, which may present substantial difficulties. Such households have, however, the option of installing their own chemical purification system for drinking water.

At present the main chemical purification option for atrazine contaminated water is treatment with granulated active carbon (GAC), an option which in the future may be combined with ozone pretreatment. This type of treatment is considered in greater detail by Nurizzo (1994). Briefly, GAC treatment can be used both for public drinking water systems and for wells supplying drinking water for just one household or for a group of households.

Operation costs for a large-scale plant (capacity about 10,000 m^3 of water per hour) are likely to be ITL 50-65 per m^3 of water. This cost takes into account e.g. energy, manpower, ordinary maintenance, GAC reactivation, and capital costs. For capital costs, at least the upper boundary of the operation costs includes depreciation of machinery and buildings. About MITL 15,000 is needed for the construction of a large-scale plant. For a small GAC plant (capacity about 50 m^3/h) installed at the outlet of a well, operation costs can be estimated to ITL 55-70/m^3 and about MITL 90 is needed for the construction. GAC treatment is likely to be successful in removing other organic contaminants, e.g. other herbicides and solvents. The presence of other contaminants may require more frequent reactivation of the GAC, but this would not incur large increases in cost.

For small wells for family supply, a reasonable choice is non-rechargeable GAC cartridges installed at the point of use. To increase the lifetime of the GAC cartridge, a small reverse osmosis plant can be used for pre-treatment. The system is only able to treat small flows of water (30-50 L/day). The costs are related to membrane and GAC replacement every 12 months and are likely to be high, about ITL 15,000-18,000/m^3. Note that only the drinking water fraction of the total water usage needs to be treated in this case. A cheaper solution may be to install a purification plant with a GAC filter for a group of households. For a water flow sufficient for about 500 persons, the operation costs can be estimated to ITL 600-700/m^3. However, relative costs per household may increase if the total water supply is treated.

Not surprisingly, it has not been possible to estimate the total costs of drinking water purification which would ensure that the EC standard for atrazine is met in Italy. Instead, this section will now focus on the situation in the area of Alto vicentino, for which data on farmers' increased costs were presented in Section 3. This increase in costs was lower than that for another maize cultivation area in northern Italy. Water purification costs in Alto vicentino are on the other hand likely to exceed the average for Italy. The reason is that the soil of Alto vicentino is mostly permeable; it is shallow and has a significant content of gravel and coarse sand. The soil permeability makes the groundwater vulnerable to pollution from the soil surface.

The aquifer of Alto vicentino contains about 3,800,000,000 m^3 of water; it provides the drinking water supply for about 180,000 persons. Every year, 360,000,000 m^3 is lost from the groundwater system to surface water by way of springs, and 58,000,000 m^3 are abstracted for public water supplies. 75-80 per cent of the consumers of water are connected to the public supply and 20-25 per cent use water from private wells.

The water purification situation vis-à-vis chemical contaminants in Alto vicentino follows closely the general description above. Thus, temporary closure of wells and blending of water are the normal measures, and chemical purification by GAC is used in emergencies. This is the case not only where herbicides contaminate water, but also for other contamination situations. For example, the two GAC plants in Alto vicentino were primarily intended to remove industrial solvents, but they are also effective for removing herbicides such as atrazine. Their combined present purification capacity is 22,075,200 m^3/year.

While it is difficult in general to estimate the costs of different routines and remedial measures (monitoring, mixing, closing wells, GAC purification, etc.) carried out to ensure that a drinking water supply meets the EC standard, it is particularly difficult to separate those costs attributable solely to atrazine contamination. The issue may be approached by identifying two cases which give boundaries for the costs. The upper boundary case, say MAX, is simply defined as a need for GAC purification of all water abstracted from the aquifer because of a total widespread contamination by atrazine. This implies a required increase in GAC plant capacity to 58,000,000 m^3/year or about 6,620 m^3/h. Leaving aside the fact that such purification will remove many other possible contaminants, and excluding water from private wells which also need to be purified, the annual cost of MAX can be estimated to be $65 \cdot 58{,}000{,}000$ = MITL 3,770.

The lower boundary case, say MIN, involves simply that part of the real water purification costs which can be attributed solely to the presence of atrazine in the aquifer of Alto vicentino, i.e.:

- Capital costs for new wells: Two new wells have been constructed as a direct consequence of the presence of atrazine in groundwater. The investment necessary probably amounted to MITL 600. Assuming a real interest rate of 6 per cent, annual capital costs amount to MITL 142.5, 81.5 and 47.0 for time scales of 5, 10 and 25 years, respectively. As the minimum cost is being sought, the 25-year figure will be used for MIN.
- Monitoring costs: About 700 samples of drinking water have been analysed for atrazine annually in the period of 1987-1991. The minimum cost per analysis is roughly estimated to be ITL 50,000, implying a minimum annual monitoring cost of MITL 35.
- Annual total costs for MIN: 47 + 35 = MITL 82.

MIN and MAX thus give the wide interval of MITL 82-3,770 for annual water purification costs to ensure that the concentration of atrazine in drinking water in Alto vicentino meets the EC standard. We now turn to a comparison between this interval and the costs of the ban in the final section.

7.5. Discussion

The MAX case of Section 4 resulted in annual water purification costs in Alto vicentino of MITL 3,770. This figure is considerably larger than the increase in the annual costs for maize farmers in Alto vicentino, MITL 483, as a result of the ban. This suggests that the annual costs of the ban at least do not exceed the upper boundary of the purification costs. The MIN case resulted in an estimate of annual purification costs of MITL 82. This is 483 - 82 = MITL 401 lower than the annual costs of a ban, but since MIN gives only the lower boundary of the purification costs, we are dealing here with an inconclusive comparison. However, the fact that the estimated costs for maize farmers are only a part of the total annual costs of the ban suggests that water purification may be the cost-effective measure.

However, within a somewhat broader perspective, the comparisons in the preceding paragraphs are misleading. Section 3 made clear that the short-term effect of restricting atrazine use is mainly a substitution by terbuthylazine and linuron. These two herbicides have leaching properties similar to those of atrazine and are thus likely to migrate to groundwater to about the same extent as atrazine. This means that the atrazine at present in the aquifer of Alto vicentino will probably be replaced by increased levels of terbuthylazine and linuron, possibly rising to concentrations that exceed the EC standard. This implies that at least the short-term effect of the ban is to merely accomplish a costly change to a mix of other contaminating substances. Comparing the costs of a ban which in fact introduces other substances in concentrations which at least in areas with permeable soils are likely to exceed the EC standard, with purification costs for atrazine alone is thus a misleading exercise.

However, the purification costs of the MAX case are still of relevance, since these are the costs of GAC treatment of all the public supply of drinking water in Alto vicentino. Such treatment would also reduce the concentrations of terbuthylazine and linuron, as well as many other organic chemicals. This means that what should really be compared with the costs of MAX are the costs for ensuring that the *total* quality of the drinking water of Alto vicentino meets the required standards. This could involve e.g. restrictions against all those herbicides whose use leads to the standard being exceeded.

In this chapter, the EC standard itself has not been questioned. Some consideration should now be given to whether this standard is warranted from the viewpoint of social efficiency. It is clear that a reduction of the concentrations of e.g. herbicides in drinking water in order to meet the EC standard involves some costs. Are there any benefits that might counterbalance such costs?

The figures above suggest that the upper boundary for the cost of ensuring that the concentration of atrazine meets the EC standard is MITL 483, i.e. the costs of a ban. If purification costs were greater than this figure, a ban would instead be the cost-effective remedial action. In Alto vicentino, there are 180,000 consumers of drinking water. Are the benefits they receive from of a drinking water quality which meets the EC standard for atrazine likely to exceed MITL 483? These benefits can be measured as the willingness to pay (WTP) per person per year for this drinking water quality. Therefore, is this WTP larger than 483/180,000 = ITL 2,683? In order to answer these questions, we must refer to the results of earlier studies.

The study that was found to be closest to this case of atrazine was that made by Mitchell and Carson (1986). Mitchell and Carson used the contingent valuation method (e.g. see Mitchell and Carson (1989) and Johansson (1993)) for estimating WTP; they asked people in an agricultural district in the United States about, *inter alia*, how they would value a very small risk reduction in cancer risk due to the presence of a chemical substance (trihalomethane) in drinking water. The estimated mean WTP, adjusted for outliers, per person per year was found to be USD 1.87 at 1990 prices, or about ITL 2,493, a figure which is very close to the ban costs per person of ITL 2,683 in the present study. As a check on the reasonableness of the WTP estimate, note that the estimate of a value of saving a statistical life (VSL) implied by the WTP of ITL 2,493 is reported to be about MITL 4,056. An extensive survey and evaluation of studies that report VSL estimates has been made by Jones-Lee (1989) (see also e.g. Viscusi (1992)). Jones-Lee concluded that it is highly probable that the VSL is at least MITL 1,250 and that it is not unlikely that it exceeds MITL 2,500. Clearly, MITL 4,056 is a large value, but it cannot be described as unreasonably large.

Although this comparison with the results of Mitchell and Carson is interesting, no clear-cut conclusion can be drawn. A major difference is that they studied a case which concerned neither atrazine nor Italy, and in contrast to the Italian case of atrazine (see Vighi and Zanin (1994)), there were actual and quantified health risks involved. Moreover, the WTP figure they obtained was for a permanent reduction of a chemical substance in drinking water, whereas the ban on atrazine is likely to result in other herbicides replacing atrazine as contaminants. However, it could be speculatively argued that the WTP figure of Mitchell and Carson can be equated with the WTP for a drinking water quality that meets the EC standard for atrazine *and* its substitutes. This would mean that if water purification is cost-effective and the costs of removing atrazine and its subsitutes are not larger than 2,493 · 180,000 = MITL 449, water purification is warrented. However, even though the WTP figure of Mitchell and Carson is for drinking water quality and not groundwater quality and hence does not include people's possible desire for pure groundwater, a ban on atrazine alone is not likely to be warrented as long as it does not result in any permanent reduction of contaminating substances in groundwater.

References

Andréasson-Gren, I-M. (1992). "Regulating the Farmers' Use of Pesticides in Sweden", Beijer Discussion Paper Series No. 10, Beijer International Institute of Ecological Economics, Stockholm.

Boadway, R. W. and N. Bruce (1984). *Welfare Economics*, Basil Blackwell, Oxford.

Baumol, W. J. and W. E. Oates (1988). *The Theory of Environmental Policy*, 2nd Edition, Cambridge University Press.

Bohm, P. and C. S. Russell (1985). "Comparative Analysis of Alternative Policy Instruments", pp. 395-460 in *Handbook of Natural Resource and Energy Economics*, Vol. I., A. V. Kneese and J. L. Sweeney (eds.), Elsevier Science Publishers, Amsterdam.

Dubgaard, A. (1987). "Taxation as a Means to Control Pesticide Application", Report No. 35, Statens Jordbrugsøkonomiske Institut, Copenhagen. (In Danish).

Eurostat (Statistical Office of the European Communitites). *Agriculture. Statistical Yearbook*, different volumes. Luxemburg.

Eurostat. *EC Agricultural Price Indices*, different volumes. Luxemburg.

Eurostat. *Yearbook of Agricultural Statistics*, different volumes. Luxemburg.

Faure, M. (1994). "The EC Directive on Drinking Water: Institutional Aspects", this volume.

Fogelfors, H., B. Johnsson, O. Pettersson and F. Petrini (1992). "Environmentally Sound Pesticide Use in Agriculture: A Discussion of Opportunities and Consequences", Report 46, Department of Economics, Swedish University of Agricultural Sciences, Uppsala. (In Swedish).

Istituto Centrale di Statistica. *Annuario di Statistica Agraria*, different volumes. Rome.

Istituto Centrale di Statistica. *Statistiche Agrarie Anno 1984*. Rome.

Istituto Centrale di Statistica. *Statistiche dell'Agricoltura, Zootecnica e Mezzi di Produzione*, different volumes. Rome.

Johansson, P-O. (1991). *An Introduction to Modern Welfare Economics*, Cambridge University Press.

Johansson, P-O. (1993). *Cost-Benefit Analysis of Environmental Change*, Cambridge University Press.

Jones-Lee, M. W. (1989). *The Economics of Safety and Physical Risk*, Basil Blackwell, Oxford.

Just, R. E., D. L. Hueth, and A. Schmitz (1982). *Applied Welfare Economics and Public Policy*, Prentice-Hall, Englewood Cliffs, N.J.

Lines, M. and M. Tamaro (1990). "Atrazine - a Non-Point Policy Problem". Working Paper 90.10, Faculty of Economics, University of Venice.

Mitchell, R. C. and R. T. Carson (1986). "Valuing Drinking Water Risk Reductions Using the Contingent Valuation Method: A Methodological Study of Risks from THM and Giardia", draft report to the U. S. Environmental Protection Agency, Resources for the Future, Washington, D.C.

Mitchell, R. C. and R. T. Carson (1989). *Using Surveys to Value Public Goods: The Contingent Valuation Method*, Resources for the Future, Washington, D.C.

North, D. (1990). *Institutions, Institutional Change and Economic Performance*, Cambridge University Press.

Opschoor, J. B. (1991). "Economic Instruments for Controlling PMPs", pp. 163-176 in *Persistent Pollutants: Economics and Policy*, J. B. Opschoor and D. W. Pearce (eds.), Kluwer Academic Publishers, Dordrecht.

Opschoor, J. B. and M. B. Vos (1989). *Economic Instruments for Environmental Protection*, OECD, Paris.

Söderqvist, T., L. Bergman, and P-O. Johansson. "An Economic Approach", chapter 12 in *Risk for Pesticides in Groundwater*, M. Vighi and E. Funari (eds.), Lewis Publishers, forthcoming in 1994.

Tietenberg, T. (1990). "Economic Instruments for Environmental Regulation", *Oxford Review of Economic Policy*, 6:17-33.

Tietenberg, T. (1992). *Environmental and Resource Economics*, 3rd Edition, Scott, Foresman and Company, Glenview, Ill.

Varian, H. R. (1992). *Microeconomic Analysis*, 3rd Edition, W. W. Norton & Company, New York.

Vighi, M. and G. Zanin (1994). "Agronomic and Ecotoxicological Aspects of Herbicide Contamination of Groundwaters Polluted by Atrazine", this volume.

Viscusi, W. K. (1992). *Fatal Tradeoffs. Public and Private Responsibilities for Risk.* Oxford University Press.

Zanin, G., M. Borin, L. Altissimo, and D. Calamari (1993). "Simulation of Herbicide Contamination of the Aquifer North of Vicenza (North-East Italy)", *Chemosphere*, 26(5):929-940.

8. CONCLUDING REMARKS AND DIRECTIONS FOR FUTURE RESEARCH

LARS BERGMAN
Professor of Economics
Stockholm School of Economics
Department of Economics
Box 6501
S-113 83 Stockholm
Sweden

MICHAEL PUGH
University College Dublin
Faculty of Veterinary Medicine
Ballsbridge
Dublin 4
Ireland

Within the frame of the Environmental Toxicology Programme of the ESF, the aim of the atrazine case study was to bring together toxicologists, institutionalists and economists for a joint analysis of a specific problem within the field of chemical safety regulation. Although the atrazine case in itself was an important one, the main concern of Working Group 4 was to use it as a tool with which reveal and clarify the major conceptual and methodological differences between the disciplines in question, and then to identify ways for fruitful cooperation. In particular a new set of interdisciplinary research projects should be identified and, where possible, pursued. The purpose of this concluding chapter is to draw some general conclusions on these issues and to indicate options for future and perhaps joint research which could lead to improvements in the field of chemical safety regulation.

Approaches to chemical safety regulation

From the beginning of the atrazine case study it was apparent that the general approach to chemical safety regulation differed considerably between toxicologists and economists. The institutionalists perhaps due to a longer history of interaction, essentially shared the view of the economists. This section aims to clarify these differences, and to explore their practical consequences in terms of the differing stringency of regulatory standards which would result.

L. Bergman and D.M. Pugh (eds.), Environmental Toxicology, Economics and Institutions, 173-181.

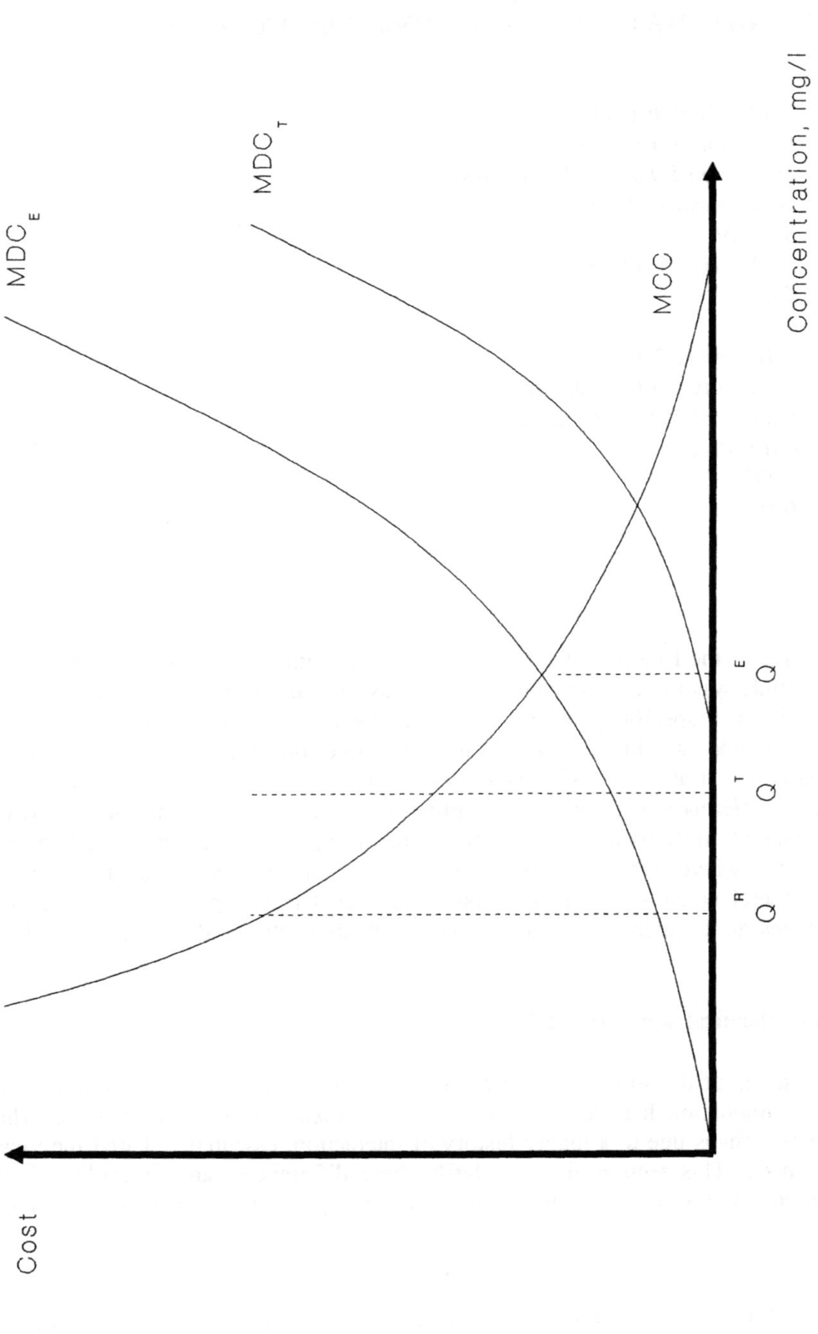

Figure 8.1 Marginal damage and control costs.

As a point of departure we sketch the conceptual framework usually adopted by economists in analyses of environmental regulation issues. This is done by means of Figure 8.1, which also reflects the main features of the atrazine case. In the figure the concentration of atrazine in drinking water is measured along the horizontal axis, while costs are measured along the vertical axis. Two types of costs are considered, and, as usual, the analysis is based on a *ceteris paribus* or everything-else-being-equal assumption.

The first type of cost is the marginal control cost (MCC), i.e. the additional cost incurred if the concentration of atrazine in drinking water is to be reduced by one unit (μg/l). The second is the marginal damage cost (MDC), i.e. the value of the damage to humans and various dimensions of environmental quality associated with an increase of the concentration of atrazine in drinking water. There are two versions of the MDC curve in the figure, MDC_T and MDC_E. The index T stands for "Toxicologists" while E stands for "Economists". Obviously these marginal cost functions are crucial in an economic analysis of the regulation of atrazine, and a few comments should be made about each of them and in particular the difference between MDC_T and MDC_E.

The MCC curve reflects the marginal cost of reducing the concentration of atrazine in drinking water. At each level of atrazine concentration the curve indicates the lowest cost for additional reductions. Thus, the curve in part reflects the cost of substituting other herbicides for atrazine, and in part reflects the cost of groundwater or drinking water purification. As it takes some time before atrazine applied for weed control ends up in drinking water, there is also a time dimension to the curve.

It follows from the assumption about cost minimization that the marginal cost of atrazine concentration reduction tends to increase as the level of atrazine concentration is reduced, i.e. the MCC curve has a negative slope. Obviously the precise shape of the MCC curve depends on the availability and cost of suitable substitutes for atrazine, as well as on the availability and cost of methods and equipment for removing atrazine from drinking water. In terms of communication between, on the one hand, economists and on the other hand, toxicologists and institutionalists, the MCC curve does not cause any particular problem. When it comes to the MDC curve, however, such problems are more apparent.

The atrazine case was selected partly because there is essentially only one type of damage that needs to be considered, namely health risks for humans. Needless to say it is a sensitive and difficult task to evaluate human health risks in monetary units. But the estimation of damage costs would be even more demanding if the use of atrazine, in addition to health effects, caused significant environmental effects. However, in the atrazine case the MDC curve only reflects human health risks due to contamination of drinking water and the known ecotoxicity of the herbicide is excluded.

Available evidence suggests that these health risks are neglible at low levels of atrazine concentration in drinking water, but may become significant at higher levels. In other words the MDC curve should have a positive slope. The level and slope of the MDC curve, however, depends on the principles for economic valuation of risks to human health. Loosely speaking the main alternatives are "objective" and "subjective" valuation. The "objective" valuation would reflect scientifically established health risks, costs for health care and productivity losses, while the "subjective" valuation would reflect perceived health risks and individual preferences concerning health risks.

As was mentioned above it early became apparent that economists and toxicologists tend to have quite different views on the basic approach to chemical safety regulation. Much of the disagreement was related to the MDC curve. In particular there were different views on the definition and valuation of the damage caused by atrazine in drinking water.

The toxicologists, believing that the only acceptable water management policy was one which ensured an absence of health risks, tended to adopt "objective" health risk valuation principles and focused on the lowest level of atrazine concentration in drinking water at which a positive value of MDC could be identified. The economists, on the other hand, adopted "subjective" health risk valuation principles and wanted to identify the value of MDC over essentially all positive levels of atrazine concentration in drinking water. As the principles for economic valuation of health risks turned out to be the key stumbling block for mutual understanding and cooperation, the main elements of the controversy need to be clarified.

From the economists point of view the empirical estimation of the MDC curve would include two phases of data collection. The first should preferably be carried out by the toxicologists and be focused on the nature and magnitude of the human health risks associated with various levels of atrazine concentration in drinking water, ranging from very low to very high levels.

These data, which reflect available scientific evidence and thus are "objective" in nature, would enable the economists to characterize the quality of drinking water quality in terms of health risks as a function of the level of atrazine contamination. It is also the type of data that the toxicologists would seek when a herbicide such as atrazine is to be regulated. But they would primarily look for the highest level of contamination at which the existence of health risks can be excluded rather than for the whole function. Moreover, the MDC_T curve in Figure 8.1 would only reflect the "objective" health risks, monetized on the basis of health care costs and productivity losses.

The next phase of the economists' data collection would be concerned with the valuation of health risks and aim at recovering individual preferences for reducing the health risks associated with atrazine contamination of drinking water. This is where the "subjective" element enters, and it is subjective because an individual may want to apply his or her own safety margins rather than the less conservative ones recommended by scientists. Individuals may be quite concerned about contamination levels well below those identified as critical on the basis of scientific evidence. In other words the existence of risk aversion suggests that the curve MDC_E in Figure 8.1 is above the curve MDC_T at all levels of atrazine concentration in drinking water.

The economists' basic reason for focusing on individual preferences in the valuation of public health risks is that the market prices of private goods, such as equipment used for purification of drinking water, do reflect individual preferences. Thus, consistent comparison of the costs and the benefits of measures to reduce the concentration of atrazine in drinking water calls for health risk valuation methods based on individual preferences.

In practise this boils down to estimating people's willingness to pay for health risk reduction.[1]

With this background the analytical framework of Figure 8.1 can be used for elucidating the differences between the approaches to chemical safety regulation usually adopted by economists, toxicologists and, in the case of atrazine, legislators. Needless to say the differences rather than the similarities between the three approaches are emphazised in the following.

From the economists' point of view regulation should be based on economic efficiency considerations. This means that the desirable level of atrazine concentration is Q^E in Figure 8.1, i.e. the level at which the marginal damage cost (MDC_E) is equal to the marginal control cost (MCC). Any deviation from this level of atrazine concentration in drinking water would imply either excessive control costs or excessive damage costs.

The toxicologists, on the other hand, would rather base the regulation on a health-based standard. That is, they would look for the maximum amount of the chemical that a human could ingest daily for a lifetime without unacceptable risk. The approach used to identify this critical amount is based on a measure of the intake of a chemical which is shown to be without unwanted effect in experimental animals. To this value is applied a safety factor of often 100 and sometimes 1000 or more in order to establish the acceptable intake for humans. In the case of atrazine the WHO standard, which is determined in accordance with the principles outlined here, is 2.0 μg/l. That is 20 times higher than the EC standard, i.e. the standard which was enforced in Italy.

In Figure 8.1 regulation of atrazine contamination in accordance with the principles generally suggested by toxicologists would correspond to a maximum atrazine concentration in drinking water equal to Q^T, i.e. a level to the left of the point where MDC_T begins to attain positive values. How much to the left of that point depends on the magnitude of the somewhat arbitrary, but widely used, safety factor mentioned above.

It should be pointed out that although Q^T is smaller than Q^E in Figure 1, this is not a necessary outcome; with a sufficiently high degree of risk aversion on the part of the affected individuals, the reverse outcome is possible. In fact, there is no *à priori* reason to expect a particular ranking. Q^T is influenced by the toxicological safety factor, and thus the degree of risk aversion on the part of the regulatory authorities, while Q^E is influenced by the degree of risk aversion on the part of the individuals being exposed to the chemical in question. Thus, the ranking of Q^T and Q^E is an empirical issue which, unfortunately, could not be suffiently explored within the frame of the atrazine case study.

The third principle is the one apparently adopted by the European legislators in the case of atrazine. It amounts to setting the maximum permissible level equal to the lowest detectable level of atrazine concentration in drinking water. This is sometimes called a "surrogate zero" value. In Figure 8.1 this level is denoted Q^R. Under all reasonable

---

[1] The theory and methods generally applied in this field of economic analysis is surveyed in Johansson et al. [1].

circumstances Q^R is less than or possibly equal to Q^T and Q^E. In other words regulation on the basis of a detection level criterion tends be quite conservative, and is more likely to produce excessive control costs than excessive damage costs.

In a dynamic perspective better measurement methods are developed, and as the detection level tends to become lower so surrogate zero moves closer toward zero. At the same time technical change and the development of new products is likely to shift the MCC curve to the left. Thus there are two types of technological change which determine the economic consequences of regulation on the basis of a detection level criterion; the rate at which measurement methods are improved and the rate at which the MCC curve shifts to the left.

Directions for future research

The atrazine case study not only revealed conceptual and methodological differences between the disciplines, but also opened up new fields of research on various aspects of chemical safety regulation. Some of these new research areas call for truly interdisciplinary approaches, while others more naturally belongs to one specific discipline. In this section some of these new research areas are briefly described following the sequence conventional to chemical development and use.

An obvious field for research which is closely related to chemical safety regulation concerns the development and marketing of new products. At a given point in time there is a given "menu" of chemical products available on the market, and this menu to a large extent determines the opportunity set of farmers and thus the cost of reducing groundwater contamination. In general the set of products supplied at a certain point in time on a certain market reflects the profit expectations held by the manufacturers some time ago. The relation between market structure and product supply is studied within the field of industrial economics. What makes this issue particularly interesting in the case of herbicides and pesticides is the interplay between technological opportunities, patent law and regulation.

For instance, the manufacturer developing a new product has to consider both the profit opportunities offered by patent protection and the potential losses due to regulatory measures during the life-time of the new product. Under certain circumstances a ban on the product just before the expiry of the patent means that it will not be available to low-price competitors. Thus contrary to common views the ban may in fact be beneficial for the manufacturer. This means that the regulator has to consider the long term effects of regulatory measures on technological change and product development.

This vision of a design strategy for the manufacturer depends on the continuation in use of command and control style legal instruments in the field of water quality, i.e. those which allow pollution to continue up to a threshold which activates a ban. The legislator has many other means of modifying manufacturer or user behaviour such as the application of a Pigouvian tax to pesticides, or offering incentives for the development of chemicals whose molecular properties would not enable them to accumulate. In effect, this would change what is presently a need for treatment circumstance to a future in which the problem could be prevented from arising. In other words it seems that analysis of the relevant incentive structures is a very important element in the development of strategies for efficient chemical safety regulation.

Beyond simple availability issues there is a need for spatially disaggregated models designed for prediction of the agricultural use of chemicals. The special challenge in the development of such prediction models is that the agricultural demand for chemicals essentially is a demand for a set of characteristics such as kill rate, cost, persistence, frequency and timing of treatment etc. Thus the demand for a specific product is derived from the demand for characteristics and depends on the properties of the product in question. It follows that a model for prediction of agricultural use of chemicals should include both a representation of the demand for characteristics and some representation of the properties of the new products that may become available during the prediction period.

Prediction of the use of chemicals, however, is only the first part of the analysis. There is also a need for empirical damage functions, i.e. models establishing a functional relation between the use of chemicals and the resulting impact on the quality of groundwater and other relevant environmental quality indicators. Lack of suitable damage functions is a general problem in empirical, policy-oriented environmental economics, and certainly a field where interdisciplinary research could be very fruitful.

Indeed, the absence of a real damage function in the atrazine case was initially seen as a difficulty by the economists in the planning stage of the project. In contrast, the toxicologists regarded the absence of damage as an indicator of a successful management policy! For toxicologists, the problem of drawing a distinction between environmental change and environmental damage remains under discussion. A third working group in the Environmental Toxicology programme of the ESF has been evaluating potential biomarkes of the environmental impacts of chemicals in different ecosystems.

One simple, but yet important, observation is that the health risks for man caused by herbicides and similar chemical substances are related primarily to their accumulated rather than instantaneous use. Further, once the concentration of the chemical had reached the critical level, the accumulation process would in general take a long time to reverse to a safe concentration level. This means that the efficiency of regulation would benefit from accurate prediction models and damage functions. With access to such tools the regulator could predict the emergence of a health risk problem and hence implement regulatory devices on the basis of ex ante rather than ex post data.

In a second component of the above mentioned Environmental Toxicology programme, models for the dispersion of chemicals through the various compartments of the environment have been under development for four years. The aim of such models is to predict the concentration of the chemical which will be achieved in the various locations and at present this is being attempted on the global level. These models may contain elements not only of exposure but also of fate, i.e. to account for what happens to the chemical after its arrival within a particular compartment. Should the substance be chemically stable in that compartment, then accumulation will be an element of its total fate. Much remains to be done in order to develop models of greater sophistication and to assess their utility in the field [1].

Another field of potentially fruitful research is related to empirical estimation of the preferences, often measured as the willingness to pay, for health risk reduction. The final aim then is to be able to empirically estimate the MDC_E curve in Figure 8.1. The basic problem is that there is no regular market for, and thus no market prices on, health risks related to drinking water contamination. Yet people do have preferences for risk reducing measures, and when these preferences are not revealed through regular market prices other methods have to be used. The available methods for estimating people's willingness to pay for health risk reduction can be grouped into the categories "direct" and "indirect" methods.

The direct method, the so called contingent valuation method (CVM), aims at revealing the individual preferences by means of a sophisticated questionaire. Although somewhat controversial CVM is widely used in empirical environmental economics, and a considerable body of methodological approaches and empirical findings have been accumulated. In the case of chemical safety regulation a CVM would be directed towards the valuation, in monetary terms, of specified health risk changes. In other words people would be asked to indicate the maximum amount of money they would be willing to pay in order to attain the specified health risk reduction.

The indirect method amounts to inferring the individual's willingness to pay for health risk reduction from his or her purchases of certain market goods. For instance, people's expenditures on safety devices in private cars reveal their preferences for avoiding the consequences of traffic accidents and the discount on private homes close to a road with heavy traffic reveals people´s preferences for silence and clean air. In the case of chemical safety regulation and drinking water, it is likely that the demand for bottled mineral water to some extent reflects the willingness to pay for avoiding contamination of drinking water. A study designed to test this proposition (with the help of Fondazione ENI Enrico Mattei, Milano) is one of the in progress activities of Working Group 4.

This concluding summary and its proposal for some alternate ways of approaching the problem of how and where to set drinking water quality standards is particularly timely as the Commission of the EU has begun a review of Directive 80/778/EEC. Among other reasons for doing so is the prevalence of pesticide exceedences of the surrogate zero threshold and, in some instances, of the WHO health-based (i.e. toxicological) standards. A real opportunity for an interdisciplinary contribution into the review process on the basis of the novel ideas which have emerged from the atrazine case study is therefore available.

Furthermore, the list of new projects on chemical safety regulation and drinking water quality can easily be extended. However, the projects already indicated are both sufficient and sufficently important to be able to conclude that close cooperation between toxicologists, institutionalists and economists is worthwhile and likely to produce relevant, practical and, above all, useful results.

References

[1] Calamari, D. (1992) *Chemical Exposure Predictions*. Lewis Publishers, CRC Press, Boca Raton, USA.

[2] Johansson, P.-O., Kriström, B. & K.-G. Mäler, eds (1994) *Current Issues in Environmental Economics*. Manchester University Press.

Economy & Environment

1. F. Archibugi and P. Nijkamp (eds.): *Economy and Ecology: Towards Sustainable Development.* 1989 ISBN 0-7923-0477-2
2. J. Bojö, K.-G. Mäler and L. Unemo: *Environment and Development: An Economic Approach.* 1990 ISBN 0-7923-0802-6
3. J. B. Opschoor and D. W. Pearce (eds.): *Persistent Pollutants: Economics and Policy.* 1991 ISBN 0-7923-1168-X
4. D.J. Kraan and R. J. in 't Veld (eds.): *Environmental Protection: Public or Private Choice.* 1991 ISBN 0-7923-1333-X
5. J.J. Krabbe and W.J.M. Heijman (eds.): *National Income and Nature: Externalities, Growth and Steady State.* 1992 ISBN 0-7923-1529-4
6. J. Bojö, K.-G. Mäler and L. Unemo: *Environment and Development: An Economic Approach* (revised edition). 1992 ISBN 0-7923-1878-1
7. T. Sterner (ed.): *Economic Policies for Sustainable Development.* 1994 ISBN 0-7923-2680-6
8. L. Bergman and D.M. Pugh (eds.): *Environmental Toxicology, Economics and Institutions.* The Atrazine Case Study. 1994 ISBN 0-7923-2986-4

KLUWER ACADEMIC PUBLISHERS – DORDRECHT / BOSTON / LONDON

9 780792 329862

MIX
Papier aus verantwortungsvollen Quellen
Paper from responsible sources
FSC® C105338

If you have any concerns about our products,
you can contact us on
ProductSafety@springernature.com

In case Publisher is established outside the EU,
the EU authorized representative is:
Springer Nature Customer Service Center GmbH
Europaplatz 3, 69115 Heidelberg, Germany

Printed by Libri Plureos GmbH
in Hamburg, Germany